T0325847

Integrating Digital Technology in Education

School-University-Community Collaboration

A Volume in Current Perspectives on
School/University/Community Research

Series Editors

R. Martin Reardon
East Carolina University
Jack Leonard
University of Massachusetts, Boston

Contemporary Perspectives on Leadership Development

R. Martin Reardon and Jack Leonard, Series Editors

Integrating Digital Technology in Education:
School-University-Community Collaboration (2019)
edited by R. Martin Reardon and Jack Leonard

Innovation and Implementation in Rural Places:
School-University-Community Collaboration in Education (2018)
edited by R. Martin Reardon and Jack Leonard

Making a Positive Impact in Rural Places: Change Agency in the Context of
School-University-Community Collaboration in Education (2018)
edited by R. Martin Reardon and Jack Leonard

Exploring the Community Impact of
Research-Practice Partnerships in Education (2017)
edited by R. Martin Reardon and Jack Leonard

Integrating Digital Technology in Education

School-University-Community Collaboration

edited by

R. Martin Reardon
College of Education at East Carolina University, Greenville, NC

Jack Leonard
University of Massachusetts Boston, MA

INFORMATION AGE PUBLISHING, INC.
Charlotte, NC • www.infoagepub.com

Library of Congress Cataloging-in-Publication Data

CIP record for this book is available from the Library of Congress
http://www.loc.gov

ISBNs: 978-1-64113-670-9 (Paperback)

978-1-64113-671-6 (Hardcover)

978-1-64113-672-3 (ebook)

Printed in the United States of America

CONTENTS

PART I

DIGITAL TECHNOLOGY IN THE ARTS

CHAPTER 1

A CURRICULAR ACTIVITY SYSTEM FOR INTEGRATING COMPUTATIONAL THINKING INTO MUSIC AND VISUAL ARTS IN THREE RURAL MIDDLE SCHOOLS

A Computer Science for All Initiative

R. Martin Reardon and Claire Davie Webb
East Carolina University

The integration of computational thinking with music and visual art aligns with the broader movement to add "arts" to upgrade the original STEM (science, technology, engineering, and math) educational focus to STEAM. We open this chapter with an acknowledgment of the sometimes fraught contexts in which curriculum is implemented. We then set forth the theoretical underpinning of the successful grant submission that funds this initiative (research-practitioner partnership and a selective overview of the concept of computational thinking leading to the adoption of an age-appropriate understanding of the concept). Other theoretical threads discussed include the model of learning that we believe is a good fit with

Integrating Digital Technology in Education:
School–University–Community Collaboration, pp. 3–29
Copyright © 2019 by Information Age Publishing

this initiative and the implementation of a conjecture map approach to developing our curricular activity system to support the integrative computational thinking learning elements. Holding professional development meetings in each of the participating districts, and valuing direct contact between us (the university faculty) and the parents of the middle school students—to the extent that this is feasible—constitute two key features of this initiative.

> *Computer Science for All is [President Obama's] bold new initiative to empower all American students from kindergarten through high school to learn computer science and be equipped with the computational thinking skills they need to be creators in the digital economy, not just consumers, and to be active citizens in our technology-driven world.* Smith (2016, https://tinyurl.com/y8d5zfpr)

Forty years ago, Duncan and Frymier (1967) compared curriculum to a soap bubble and sagely observed that an analysis of a soap bubble may miss the point. They facetiously cautioned that such an analysis had the potential to result in merely "a conceptual wet spot on [the analyst's] intellectual fingers" (p. 180). We would prefer to compare the curricular activity system that we discuss in this chapter to something far more substantial than a soap bubble, but we heeded their word of caution and have taken care that our analysis does not leave us with sticky intellectual fingers.

The supportive conditions one would wish to establish for our fledgling curriculum initiative include that it is being implemented in the context of a robust research-practice partnership under the leadership of three supportive principals by three highly competent and involved teacher partners. The conditions that detracted from the implementation of the initiative in the first full year included two hurricanes that deluged the whole region with about three feet of rain, led to record-setting flooding, destroyed the houses of two of our teacher partners, resulted in students missing almost 7 weeks of school, and prompted the decision to demolish one of the middle schools in which our curricular activity system was being implemented.

Against this backdrop, we return to the point of this chapter. The *Computer Science for All* (CS4All) initiative was established under the auspices of President Obama in 2016 to focus on promoting the learning of computer science by students in all grade levels in all schools. The distinction between computer science and computational thinking that is evident in the epigraph of this chapter clearly existed from the outset. We explore the nuances of computational thinking as a term later in this chapter, but, at the outset, it conjures up its role in mathematics- and science-related fields—in addition to an obvious association with computer science-related fields.

By contrast, we begin by discussing the genesis and early implementation of a National Science Foundation (NSF) grant-funded endeavor to integrate computational thinking into the teaching of general music and visual arts in three rural middle schools in eastern North Carolina. To clearly associate our endeavor with CS4All, following an early suggestion of one of our collaborative partners, the acronym we adopted for "Integrating the Computer Science for All Initiative in Three Rural Eastern North Carolina School Districts" was "iCS4All." In a welcome and generative departure from what might have been expected, the focus of our endeavor is the seamless integration of computational thinking with the North Carolina state-defined standards in the general music and visual arts subject areas—specifically at the Grade 8 level.

Research-Practitioner Partnerships

In the solicitation to which our iCS4All grant proposal responded, NSF stressed the imperative for proposals to emerge from research-practice partnerships (RPPs, https://www.nsf.gov/funding/pgm_summ.jsp?pims_id=505359). RPPs were characterized by Coburn, Penuel, and Geil (2013) as "long-term, mutualistic collaborations between practitioners and researchers that are intentionally organized to investigate problems of practice and solutions for improving district outcomes" (p. 2). Embellishing this concept, Penuel, Allen, Coburn, and Farrell (2015) drew a distinction between partnerships that are oriented to the translation of research into practice on the one hand and RPPs on the other, proposing that RPPs constitute "a form of joint work requiring mutual engagement across multiple boundaries" (p. 182). RPPs, Penuel et al. claimed, enhanced the likelihood of the desirable educational outcomes previously chronicled by Fishman, Penuel, Allen, and Cheng (2013) and subsequently discussed by Coburn and Penuel (2016).

In this instance, iCS4All was built on an existing and still ongoing RPP initiated described by Militello, Jones, Dunn, and Marshburn Moffitt (2018), that is funded by the Panasonic Foundation. This preexisting RPP continues to focus on improving teacher quality in four rural school districts, three of which were keen to extend that preexisting collaboration to embrace iCS4All. The understanding and trust initially engendered by the preexisting RPP has contributed substantially to the effectiveness of our integrative approach. To provide just one instance—the process for purchasing materials to be used by the teachers needed to be clearly laid out. After discussions between the principal investigator and the district financial managers, decisions about the purchase of materials are authorized at a moderation meeting (see later in this chapter) between the university faculty involved in iCS4All and the teachers and then reauthorized by each school's principal before being processed through each individual school district's process. When the materials are delivered, invoices are forwarded to East Carolina University (ECU).

ECU's central grant administration before being forwarded to iCS4All's financial administrator who authorizes reimbursement from the grant funds. We mention this because it is the sort of infrastructural departure from "business as usual" for the school districts that could have been problematic in the absence of the level of an established level of understanding and trust among our school district partners and ourselves.

FROM STEM TO STEAM

The 5-year strategic plan for science, technology, engineering, and mathematics education (STEM; National Science and Technology Council, 2013) asserted that the "jobs of the future are STEM jobs," and that "progress on STEM is critical to building a just and inclusive society" (pp. vi–vii). This ringing endorsement of the pivotal role of STEM education overlooked what Robelen (2011) had earlier construed as momentum for exploring perspectives from the arts to enrich student engagement in STEM fields— momentum that led to the amending of the STEM acronym by likeminded stakeholders to STEAM by adding an "A" for "arts" (see also http://stemtosteam.org; Lachman, 2018; Sousa & Pilecki, 2013).

In iCS4All, we are reversing the conventional directionality of bringing arts to STEM by bringing computational thinking to music and the visual arts in the context of our school-university-community collaboration. In the true spirit of an RPP—which envisages that the practitioner partners should play a major role in determining the focus of the research—the initiative to focus on music and visual arts came from the school district partners. The district representatives from the three districts that were keen to be involved were invited to attend the first planning meeting for developing a response to the NSF call for proposals and were asked to consult within their districts beforehand and bring their best thinking about how they would like to see the grant implemented, should it be funded. In addition, it was suggested that the district representatives might go to the extent of canvassing their respective districts for teachers who might be interested. To the surprise of some at that first planning meeting, two of the districts proposed that their highly creative and innovative visual arts teachers be involved, and the third district proposed that its similarly highly creative and innovative music teacher (who had earned her doctorate as a music educator) be involved. That NSF added iCS4All to the numerous other NSF grants that are oriented to the integration of computational thinking with the arts is a validation of the reorientation of STEM to STEAM.

COMPUTATIONAL THINKING: AN OVERVIEW

There is an abundance of definitions of computational thinking through which our overview traces a selective path. Wing's (2006) frequently-cited succinct *Viewpoint* article for the Association for Computing Machinery (ACM) laid the foundation for much subsequent discussion. Wing highlighted aspects of computational thinking in action such as when someone is solving problems, designing systems, or understanding human behavior. Wing envisaged computational thinking as, for example, reformulating an insuperable problem to open the way for reduction, embedding, transformation, or simulation, and as thinking recursively, utilizing parallel processing, and type checking, or using abstraction or decomposition. While giving her credit for bringing the term computational thinking into the limelight, there are some who consider Wing's advocacy as initiating a rash of imprecision, as we will discuss shortly.

Four years later, the Committee for the Workshops on Computational Thinking of the National Research Council (NRC) convened the first of two workshops to glean insights into computational thinking of invited "education researchers and cognitive scientists familiar with [its] educational dimensions" (NRC, 2010, p. 1). The highly qualified participants in the first National Research Council (NRC, 2010) workshop—of which Wing (2006) was also a member—contemplated avoiding the definitional issue by embracing computational thinking as "the union of [multiple] views—a laundry list of different characteristics" (NRC, 2010, p. 59). Consequently, the participants endorsed a laundry list, proposing that computational thinking

> might include reformulation of difficult problems by reduction and transformation; approximate solutions; parallel processing; type checking and model checking as generalizations of dimensional analysis; problem abstraction and decomposition; problem representation; modularization; error prevention, testing, debugging, recovery, and correction; damage containment; simulation; heuristic reasoning; planning, learning, and scheduling in the presence of uncertainty; search strategies; analysis of the computational complexity of algorithms and processes; and balancing computational costs against other design criteria. Concepts from computer science such as algorithm, process, state machine, task specification, formal correctness of solutions, machine learning, recursion, pipelining, and optimization also find broad applicability. (NRC, 2010, p. 3)

Describing computational thinking as a laundry list is somewhat pejorative, but it highlights the elusiveness of the concept. At the second NRC workshop a year later, the participants considered not defining the term at all before determining that to do so would be "deeply unsatisfying" (NRC, 2011, p. 3). The difficulty in listing such a wide range of practical applications and situations in which computational thinking plays a role is that it leaves it up to the reader to distill his or her generalized understanding from a wealth of highly esoteric instances with little guidance as to the accuracy of his or her formulation. In Aho's (2011) contribution to a symposium sponsored by the ACM, he asserted that "we consider" computational thinking to be "the thought processes involved in formulating problems so their solutions can be represented as computational steps and algorithms" (p. 2). Aho began his paper by lauding clarity of definition and precise specification—arguably in reaction to the laundry list approach—but it is unclear to whom Aho was referring when he chose to use the plural "we." Further, as will be discussed shortly, even the reference to thought processes has been viewed subsequently as problematic by Denning (2017b).

In 2016, on the 10th anniversary of her 2006 *Viewpoint* article on computational thinking, Wing (2016) highlighted the extent to which computer science had been adopted into the K–12 curriculum but refrained from further addressing the definition of computational thinking. Adopting a broader perspective, Wing declared that "computational thinking will be a fundamental skill used by everyone in the world by the middle of the 21st century" (para. 7). Wing's enthusiasm has not been universally endorsed. For example, Denning (2017b) recently took exception to "unsubstantiated claims promoted by enthusiasts" (p. 34) regarding the ubiquity of computational thinking and blamed such enthusiasts for the dilution of what he regarded as initially clear definitions of computational thinking to the stage where the concept has become "vague and confusing" (Denning, 2017b, p. 33).

Denning (2017b) traced the "rich pedigree" (p. 34) of the term computational thinking from George Polya in the 1940s, through other leading figures in the field such as Alan Perlis, Allen Newell, and Herbert Simon in 1960s, to Donald Knuth in the mid-1970s and Edsger Dijkstra in the late 1970s. Denning conjectured that Seymour Papert may have been the first to use the term "computational thinking" in *Mindstorms* (1980) to describe children's thinking as they programed with LOGO. Denning (2017a, 2017b) highlighted the emergence of the term from a more rarified provenance, not in the field of computer science itself, but from physics in the early 1980s under the influence of Nobel Laureate in physics, Ken Wilson.

Writing about the role of computational thinking in science, Denning (2017a) evoked the concept of "a quiet but profound revolution" (p. 13) permeating scientific fields. He paraphrased what he referred to collectively as "most published definitions" of computational thinking as "the thought processes involved in formulating problems so that their solutions are represented as computational steps and algorithms that can be effectively carried out by an information-processing agent" (p. 3). He then proceeded to characterize as problematic "fuzzy thinking" (p. 3), the references to "formulating," "information agent," and "thought processes" (p. 3). As the laundry list approach demonstrated, it is easier to cite instances of computational thinking than it is to define it. For example, Denning (2017b) compared what he referred to as "traditional computational thinking" and "new computational thinking" as shown in Table 1.1.

Table 1.1.

Comparative Description and Instances of Traditional and New Computational Thinking

Traditional Computational Thinking		New Computational Thinking
Mental habits and disciplines for designing useful software	1	Formulating problems so that their solutions can be expressed as computational steps
Extensively practicing programming cultivates computational thinking as a skill set	2	Computational thinking is a conceptual framework that enables programming
Skills of design and software crafting— for example, separation of concerns, effective use of abstraction, devising notations tailored to one's needs, and avoiding combinatorically exploding case analyses	3	*Set of problem solving concepts* such as representation, divide-and-conquer, abstraction, information hiding, verification, and logical reasoning
A new way of conducting science, alongside theory and experiment—a revolution in science	4	*Useful in* sciences and *most other fields*
Algorithms are directions to control a computational model (abstract machine) to perform a task	5	*Algorithms are expressions of recipes for carrying out tasks;* no awareness of computational models is needed
Programs are tightly coupled with algorithms; programs are algorithms expressed in a computer language; algorithms derive their precision from a computational model	6	Programs are loosely coupled with algorithms; *algorithms are for all kinds of information processors, including humans*—it is completely optional whether an algorithm will ever be translated into a program
Designing computations in a domain requires extensive domain knowledge	7	*Someone schooled in the principles of computational thinking can find computational solutions to problems in any domain*
End users can follow algorithms and get the result without any understanding of the mechanism	8	*People engaging in any step-by-step procedure are performing algorithms and are* (perhaps unconsciously) *thinking computationally*
Engaging in a computational task without awareness is not computational thinking	9	*People who are engaging in any task that could be performed computationally* are engaging in subconscious *computational thinking*

Note. Adapted from "Remaining Trouble Spots with Computational Thinking," by P. J. Denning, 2017, *Communications of the ACM, 60*(6), p. 37. Copyright 2017 by the Association for Computing Machinery. Italics in the New Computational Thinking column added.

Denning's (2017b) assertion in Table 1.1 of the role of subconscious/unconscious processes in computational thinking and his assertion that computational thinking is independent of any computational model or programing—even though it facilitates programing—is striking. We will return to this point later, but at this juncture we advert to the consonance between elements of Denning's description of new computational thinking (highlighted by the added italics in numbers 3 through 9) and the United Kingdom's Computing at School's (n.d.) CAS Barefoot conceptualization of computational thinking that we adopted for iCS4All.

The CAS Barefoot project was established in 2014 to render assistance to primary school teachers (who teach students up to 11 years-of-age) who were preparing to implement a new computing curriculum. Now led and funded by BT (formerly British Telecom), and an integral part of the Computing at School (CAS) initiative (http://www.computingatschool.org.uk/), CAS Barefoot conceptualized computational thinking as consisting of six concepts—closely aligned with the italicized elements of Denning's (2017b) new computational thinking in Table 1.1—and five approaches. As shown in Figure 1.1, the six concepts include establishing and checking facts (logic), facility with precise sequences of instructions or sets of rules (algorithms), ability to break a problem or a system into parts (decomposition), making predictions and creating rules (patterns), identifying essentials (abstraction), and making judgements based on pertinent factors such as design criteria (evaluation). In the process of mastering the six concepts, CAS Barefoot anticipated that the primary school learner would employ one or more of five approaches that include tinkering, creating, debugging, persevering, and collaborating.

CAS Barefoot's conceptualization of computational thinking was developed by a group of highly skilled K–12 educators who were seeking to enunciate an understanding that would be pertinent to practice in schools (J. Waite, personal communication, 2018,) rather than as an outcome of a higher education-oriented approach to formal definition. We were attracted to the operational nature of this CAS Barefoot understanding of computational thinking, by the proximity of the CAS Barefoot students to the age level of the iCS4All Grade 8 middle school students, and by how easily the concepts and approaches framed the implementation of iCS4All in the fields of music and visual arts. As validation of our sense of "goodness of fit," after an initial introduction to the CAS Barefoot computational thinking concepts, the music teacher commented with conviction that "the six CAS Barefoot concepts put words on what we do all the time in music education."

Broadening the Context of Computational Thinking

Having clarified that our project is firmly rooted in the new understanding of computational thinking (Denning, 2017b) as encapsulated

Source: Reproduced from https://barefootcas.org.uk/computation-thinking-poster/
(Reproduced with permission.)

Figure 1.1.   Graphic depiction of the concepts and approaches associated with computational thinking.

in the CAS Barefoot formulation of concepts and approaches, we note that the new understanding explored by Denning (2017b) had been previously ascribed a role commensurate with that ascribed to STEM by the National Science and Technology Council (2013). Previously again, although the NRC (2010) workshop participants envisaged the field of computer science as providing the theoretical underpinning and the terminology to describe the notions associated with computational thinking, their unanimous understanding was that computer scientists were not the only individuals who engaged in computational thinking. To the contrary, they unanimously asserted that "computational thinking is comparable in importance and significance to the mathematical, linguistic, and logical reasoning that society today agrees should be taught to all children" (p. 3).

Computer science, in the NRC (2010) workshop participants' understanding, provided a framework and a vocabulary that grounded computational thinking from the perspective of "automating computational abstractions" (p. 7), but computational thinking "is likely to benefit not only other scientists but also everyone else—bankers, stockbrokers, lawyers, car mechanics, salespeople, health care professionals, [and] artists" (p. vii). Indeed, numerous NRC workshop participants supported the idea that computational thinking "could be better understood as a fundamental intellectual skill comparable to reading, writing, speaking, and arithmetic" (p. 13).

EDUCATION AND COMPUTATIONAL THINKING

The confidence of the participants in the NRC (2010, 2011) workshops regarding the central role in education of computational thinking echoed the earlier confidence of Wing (2006)—herself a participant in those workshops—who proposed multiple ways of conceptualizing computation thinking. The NRC (2010) workshop participants highlighted five reasons for the educational importance which they ascribed to computational thinking: (a) equitably equipping students to access current technological resources, (b) arousing interest among students in careers in information technology, (c) maintaining the economic competitiveness of the United States—echoing the urgency conveyed by the National Academy of Sciences, National Academy of Engineering, and Institute of Medicine (2007) in their call to arms, (d) facilitating students' engagement in other disciplines, and—closely related to the initial reason—(e) empowering individual students to efficiently and effectively negotiate contemporary society.

With respect to this final reason, Pea (one of the participants in the NRC 2010 workshop) declared that computational thinking was implicit in everyday life—that, echoing the assertion of the iCS4All music teacher quoted above, it was a way of describing how people already go about their business. From Pea's perspective, "connecting computational thinking in a personally meaningful way is at the heart of tackling the problem of how everyone can be brought into a pathway for developing and using [computational thinking] in their everyday lives" (NRC, 2010, p. 6).

To hearken back to the second of the two workshops conducted by the Committee for the Workshops on Computational Thinking (NRC, 2011), the issue of how to forge personally meaningful connections between computational thinking and individuals' everyday lives was carefully considered. Of the seven pedagogically relevant questions posed to the participants, our partial paraphrase of the two that are most pertinent in the context of iCS4All are:

- How can computational thinking be integrated most effectively in other subjects taught in the classroom?
- How should we measure the success of efforts to teach computational thinking? (p. 2)

Integrating Computational Thinking With Other Subjects

With respect to the first of our paraphrased questions, the NRC (2011) workshop participants explored a wide range of contexts in which computational thinking plays a clear role. These included everyday life contexts such as troubleshooting digital devices (perhaps by turning them off to return them to a default state), or by choosing which supermarket checkout line to join. NRC workshop participants also cited numerous digital games that favor the computational thinker, and, in the formal educational realm, they expressed their agreement that computational thinking underpinned the learning of numerous subfields of science, engineering, journalism, and medicine.

Widening the field of view, Yadav, Hong, and Stephenson (2016) urged participation from the instructional technology community to influence vision and action by "helping to articulate the connection between computational thinking and all academic disciplines, developing content to support integration into curricula, and taking the lead in designing and facilitating both preservice and inservice opportunities for learning" (p. 568). By their participation, Yadav et al. suggested, advocates for computational thinking integration can be catalysts of conversation "with and between leaders in all academic subject areas" (p. 568).

iCS4ALL IN ACTION

We commenced being catalysts of conversation along the lines of the Yadav et al. (2016) suggestion when iCS4All was funded by NSF on October 1, 2018. iCS4All is a collaboration between two departments at East Carolina University (Educational Leadership and Computer Science), The William & Ida Friday Institute for Educational Innovation at North Carolina State University, and three rural eastern North Carolina school districts that were partners in a pre-existing and ongoing research-practitioner partnership.

As shown in Figure 1.2, iCS4All was designed to engage students, teachers, and principals as well as parents. We consider that professional development for both the principals and the teachers is crucial to the successful integration of our computational thinking curricular activity system (Roschelle, Knudsen, & Hegedus, 2010; see subsequent discussion) with the teaching of music and visual arts. Equally crucial is the involvement of the parents of the students involved in iCS4All.

In the absence of collaboration from the school district and the input of the principals and teachers, the schematic in Figure 1.2 would have remained just that. An example of the level of commitment from one of the principals will serve as an illustration of the reality of the support the

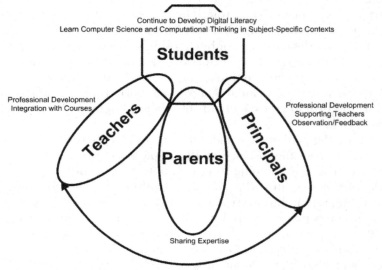

Figure 1.2.   Conceptual Schema for iCS4All.

collaboration has enjoyed. At one of the professional development events organized for the teachers and principals (design-based implementation research; see later), one of the principals received a text at lunch time informing him of an unanticipated crucial meeting to be held that night in his home town. Following the after-lunch session, he drove 3 hours to attend that meeting, stayed at his house overnight, and then drove back to join his teachers for the second day—arriving for the breakfast session.

We believe that reaching out to parents in the first place and intentionally engaging them in a conversation about the purpose of iCS4All and the elements of the curricular activity system we have devised are key features of iCS4All. Rural education researchers have highlighted the mixed feelings with which some parents in some rural communities wrestle regarding education. Naturally, parents want the best for their children, but while Reardon (2018) warned against over simplification and comprehensively nuanced the understanding, he drew attention to education as a factor in the "hollowing out of America's rural places" (p. xiii; see also Carr & Kefalas, 2009). In a more acerbic formulation of the issue, Corbett (2007) went so far as to refer to schooling in rural places as the "quintessential institution of disembedding" (p. 251). For our part, we see the increasing penetration of digital technologies into the agricultural sector as potentially making employment opportunities available in rural places for highly skilled workers, and the potential for highly educated individuals to telecommute to work thereby making it feasible for them to live in rural communities

Developing the iCS4All Curricular Activity System

According to Resnick (NRC, 2011, pp. 67–69), the emergence of students' ability to meaningfully express their ideas depends on their development of both their concepts (i.e., computational thinking, in the case of iCS4All) and their capacities. Resnick went on to conjecture that students' capacities to design solutions that effectively cope with the complexity and interactivity of contemporary society thrive in the synergistic affordances of a socially cooperative learning environment.

How can computational thinking be integrated most effectively in music and visual arts? Our extended answer to this key question—the first of the two most pertinent questions raised by the Committee for the Workshops on Computational Thinking (NRC, 2011)—begins with the work of Roschelle et al. (2010) who proposed an approach to curriculum design that conformed to Resnick's (NRC, 2011) conjecture. They initially explored the dual meanings of the word "curriculum" as connoting either "a framework of teaching objectives or a specific textbook that fulfills such a framework" (p. 239). In terms of curriculum as a framework of teaching objectives, they developed the concept of a curricular activity system, in which both a framework of objectives and the fulfillment of this framework coexisted. Each of the terms in "curricular activity system" was meaningful. For example, they chose the word "curricular" to portray the importance of progress in student ability across the hours of their engagement with instruction. The imperative for active participation on the part of the student is underscored by the term "activity"—emphasizing that learning will not take place because of lessons or problems, but through activities that require the student to be engaged and creative. Further, this active participation is intentional and characterized "in terms of its objective (for the participants), available materials, the intended use of tools, the roles of different participants, and the key things we would like participants to do and notice" (p. 239). Finally, Roschelle et al. referred to their concept of curricular activity as a "system" because the incorporation of each component such as the use of technology or the provision of teacher professional development is specifically oriented to generating an impetus for students to more closely approach the learning objective.

The Roschelle et al. (2010) curricular activity system approach is particularly well-suited to iCS4All because it highlights both "knowledge building and learning progressions" (p. 240). In integrating computational thinking in creative subject areas such as visual arts and music, our intention is to collaborate with the teachers in the three districts to build the middle school students' knowledge of the six concepts of computational thinking (see Figure 1.1). Before discussing the elements of our curricular activity system in detail, we will discuss some associated aspects of iCS4All that will set our endeavor in context.

How should we measure the success of our efforts to teach computational thinking? In response to this second of the two most pertinent questions raised by the Committee for the Workshops on Computational Thinking (NRC, 2011), we gauge the success of our efforts in iCS4All by tracking the progress of the middle school students' learning through (a) participant observation in the classroom, (b) students' self-assessment of the learning approaches they implement in engaging with the iCS4All curricular activity system elements, (c) reviewing artifacts that students produce as they engage with the iCS4All curricular activity system elements, (d) conducting focus groups of students, and (e) anonymous snapshot "exit assessment" surveys.

Moderation Meetings

With respect to (c) above, the teachers who are participating with us in iCS4All meet monthly at each of the sites in turn with a faculty member from educational leadership and faculty member from computer science both of whom are committed to the success of iCS4All. These meetings are referred to as "Moderation Meetings." One of the functions of Moderation Meetings is to provide a venue in which each teacher (a) shares artifacts that his or her students produced in response to the curricular activity system element over the course of the prior month and (b) explains in the context of a modified tuning protocol (McDonald & Allen, n.d.) setting how he or she construes these artifacts as supporting his or her claim that the students employed computational thinking.

The teachers also preview the curricular activity system element scheduled for the following month and receive approval for the purchase of equipment to implement that element. For example, at one Moderation Meeting the music teacher received approval to purchase 30 iPads, and each of the visual arts teachers received approval to purchase five desktop computers.

THE CONNECTION BETWEEN COMPUTATIONAL THINKING AND CREATIVITY

As alluded to earlier, though commonly associated with computer science, the fact that computational thinking is a distinct capacity opens the door to discussing its role in multiple avenues of human creativity (Mishra & Yadav, 2013). Conceptualizing computational thinking as a process of problem-solving engaged in by a human who is using a computer as a tool in developing a creative response to the problem facilitates a wide range of applications across many disciplines. In this sense, taking a computational thinking approach eschews the complete reliance on computers and technology to solve a problem, but instead encourages students' critical and creative thinking to sharpen their ability to ask questions, learn from failed attempts, and create new ways to overcome obstacles to achieving the desired outcome. Therefore, one could argue that computational thinking is a process that is most securely anchored in creativity—even in computer science contexts. According to Mishra and Yadav (2013), "computational thinking can foster creativity by allowing students ... not only [to] be consumers of technology, but also [to] build tools that can have significant impact on society" (p. 11). By empowering students to take ownership of what they are creating and learning, a deeper level of engagement across disciplines is enabled. Mishra and Yadav argued that new possibilities emerge when computational thinking and a "deep knowledge of a discipline" (p. 11) are combined. This deep knowledge is an irreplaceable part in the creative process that allows for collaboration between human and technology in creative productivity.

Yadav et al. (2016) explored challenges inherent in some rural settings when it comes to connecting computational thinking and creativity. In an era when standardized testing outcomes contribute to the evaluation of quality of a school, it can be a challenge to establish the value of an emphasis on creativity in teaching when creativity is not oriented to making an impact on standardized test scores. Further, in the absence of grant funding like iCS4All, it can be a challenge to find a way to implement a learning approach the foregrounds computational thinking while also "operating within the constraints of available resources" (p. 565).

Since much research about computational thinking is focused on its applicability across disciplines, schools can bring computational thinking into a range of already existing courses such as art, social studies, music, and writing even in the absence of computing facilities. For example, Yadav et al. (2016) discussed teaching students algorithms as problem-solving tools by deconstructing tasks into steps to be taken to reach an outcome, or by inviting students to analyze activities consisting of a sequence of steps such as baking a cake or brushing their teeth. A similar process can be adopted with abstraction by helping students to detect the essential elements of a solution to one problem and apply those elements in a different context.

Integrating Computational Thinking Into Music and the Visual Arts: Two Examples

A thought-provoking example of the integration of computational thinking into music was discussed by Mishra and Yadav (2013). A computer program, EMI (referred to as Emmy), was created by musician and composer David Cope, and "was once the world's most advanced artificially intelligent composer" (Blitstein, 2010). Cope used Emmy to produce computer-generated music that not even the most sophisticated listeners could recognize as machine-made (Mishra & Yadav, 2013), including compositions in the style of famous composers. Blitstein (2010) noted that Cope's work was regarded with awe but also with some reservation—especially when questions were raised about who should take credit for the music: the person or the computer. However, for Cope, Emmy was only a tool in his creative hand. Cope's response predated a contemporary understanding that "computational thinking and creative thinking are ... separate but compatible cognitive tools that individually expand the ways knowledge and skills can be applied to problem-solving situations" (Peteranetz, Flanigan, Shell, & Leen-Kiat, 2017, p. 305).

In the field of visual arts, Knochel and Patton (2015) proclaimed the value of students' integration of computational thinking with creativity to develop their own code to produce art and staunchly argued that writing code and utilizing technology in art is not simply applying mathematical algorithms "but rather a process of design, an act of free speech, and a digital production method" (p. 34). Recently, they declared, art education programs have been urging students to improve their digital literacy as well as their knowledge about how society is shaped by digital media. They asserted that "computational thinking through critical digital making should be embedded in the art curriculum to equip students with coding skills important to digital literacy" (p. 24). Knochel and Patton provided a striking example of computational thinking in art by citing Lozano-Hemmer's *Tape Recorders* (2011)—a gallery installation in which

> sensors in the gallery space recognize where viewers stand, activating tape measures to extend out, symbolically representing the length of time audience members linger in front of the artwork and creating a physical manifestation of the institutional space of art engagement. Lozano-Hemmer's software exposes a physical metric of museum visitation, raising questions around how we quantify the quality of engagement with artworks. (p. 32)

MODEL OF LEARNING

The last of the associated aspects of iCS4All that will set our endeavor in context before we return to a discussion of the elements of our curricular activity system is the model of learning—distilled from a synthesis of many meta-analyses by Hattie and Donoghue (2016, 2018)—that supports our endeavor. We are aware of the skepticism of some who have referred to Hattie's distillation of meta-analyses as pseudoscience (Bergeron, 2016) or who have critiqued his exaltation by some to the status of guru as a harbinger of a neo-Taylorist era in education (Eacott, 2017); however, our choice of the Hattie and Donoghue (2016) model of learning is pragmatic and based on its fit with our context. They described a model of learning that consists of

> three inputs and three outcomes; student knowledge of the success criteria for the task; three phases of the learning process (surface, deep and transfer), with surface and deep learning each comprising an acquisition phase and a consolidation phase; and an environment for the learning. (pp. 1–2)

As shown in Figure 1.3, the three inputs are the same as the three outcomes and are referred to as skill (the student's prior or subsequent achievement), will (the student's learning disposition), and thrill (the range of motivations experienced by the student). Effective learning effectively promotes greater skill, will, or thrill, with all three inputs/outcomes valued equally. The student's knowledge of the success criteria in a prelearning phase orients his or her efforts. Then, acquiring surface learning (through strategies like learning vocabulary or note taking) supports subsequent consolidation

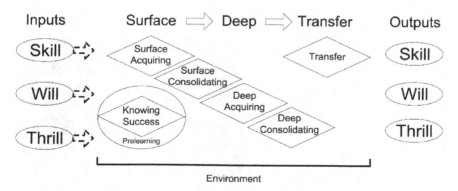

Source: Adapted from "Learning Strategies: A Synthesis and Conceptual Model," by John A. C. Hattie and M. Donoghue, 2016, *NPJ Science of Learning,* p. 2. Copyright the authors.

Figure 1.3.   Model of learning.

of the surface learning (through strategies like learning how to benefit from feedback). At some juncture, sufficient surface learning supports the acquisition phase of deep learning (through strategies like elaboration and organization and strategy monitoring) and the acquisition phase supports the consolidation phase of deep learning (through strategies like self-verbalizing the steps in a problem or seeking help from peers and peer tutoring). While this model of learning does not postulate a strict hierarchy of learning the governing principle is that a learner cannot move straight to higher level learning without a sufficient grounding of lower level learning. Eventually, a learner may be able to transfer his or her knowledge and understanding from one situation to an analogous situation (near transfer) or even to a completely different situation (far transfer).

iCS4ALL CURRICULAR ACTIVITY SYSTEM

All of the above supported our construction of the curricular activity system we designed for iCS4All. We found the conjecture mapping approach to systematic educational design research (Sandoval, 2014) assisted us in conceptualizing the elements of the system—particularly the "if . . . then" formulation of the design conjectures and the theoretical conjectures, and conceptualization of activity structures as incorporating both task structures and participant structures (see Figure 1.4).

We designed a 7 month-long set of curricular activity system elements for both music and visual arts, culminating with an eighth "portfolio" element that prompted the students to accumulate the artifacts they produced

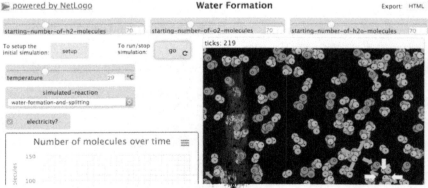

Source: Adapted from "Conjecture Mapping: An Approach to Systematic Educational Design Research," by W. Sandoval, 2014, The Journal of the Learning Sciences, 23, p. 21. Copyright 2014 by the Taylor & Francis Group, LLC.

Figure 1.4. Generalized conjecture map.

throughout the year for an exposition to coincide with the final meeting for the year of their school's parent-teacher organizations. Each element was designed to align with one of the two most pertinent North Carolina Grade 8 Essential Standards in each of the subject areas as shown in Table 1.2.

To bring our chapter to a close, we will discuss a music and visual arts curricular assessment system element in terms of their activity structures, our corresponding design and theoretical conjectures (Sandoval, 2014), and the computational thinking concepts we have targeted. We made a considered decision to refrain from external evaluation of the approaches that the students implement both because of the highly inferential nature of such judgements and because collecting data to support such judgements

Table 1.2.

Alignment of Curricular Activity System Elements With Pertinent North Carolina Essential Standards

Music	
Essential Standard 8.ML.2	M1: Musical Instrument Digital Interface (MIDI) and Composer's Cut Video Composition
Interpret the sound and symbol systems of music	
Clarifying Objective 8.ML.2.1	
Interpret standard musical notation for whole, half, quarter, eighth, sixteenth, and dotted note and rest durations in 2/4, 3/4, 4/4, 6/8, 3/8, and alla breve meter signatures	M2: MIDI Composition Extension
Essential Standard 8.ML.3	M3: Sound Montage
Create music using a variety of sound and notational sources.	M4: Teach Me CT
Clarifying Objective 8.ML.3.2	M5: Sounds of Home
Construct short pieces within specified guidelines . . . using a variety of traditional and non-traditional sound, notational and 21st century technological sources	M6: Multiliteracies M7: Create Your Own Instrument
Visual Arts	
Essential Standard 8.V.2 Apply creative and critical thinking skills to artistic expression.	V1: Palimpsest
Clarifying Objective 8.V.2.1 Create art that uses the best solutions to identified problems.	V2: Set
Essential Standard 8.V.3 Create art using a variety of tools, media, and processes, safely and appropriately.	V3: Digital Mondrian V4: Multiliteracies
Clarifying Objective 8.V.3.2 Use a variety of media to create art (digital media are referenced at the Grade 6 level).	V5: Composer's Cut Video V6: Sights of Home V7: Graffiti Art

validly and reliably is excessively resource-intensive in this context. All students will be asked to self-assess and report the intensity with which they adopted each of the five approaches when each element draws to a close on a continuous Likert scale ranging from "min." to "max."

M1: Musical Instrument Digital Interface (MIDI) and Composer's Cut Video Composition

In this opening element of the music curricular activity system, students are firstly invited to collaborate with a colleague to conduct their own Internet research to explore a different "symbol system of music" (Essential Standard 8.ML.2) and teach each other about it. The extent of their learning and comprehension will be reflected in their ability to use that new symbol system to accurately copy eight bars of a tune of their own choice. They are then challenged to compose an eight-bar tune of their own and, finally, to self-assess their learning by composing a short video that features their own composition. We designed this and all the other elements to be implicitly open-ended. For example, we anticipated that students with the appropriate "skill" and "will" may compose well beyond eight bars for the "thrill" of it (Hattie & Donoghue, 2016). As a full example, Table 1.3 is a compilation of the features of the M1 element.

V1: Palimpsest

While we did not place a premium on paralleling the focus of the elements between the music and visual arts domains, some direct comparisons emerged as our design work proceeded. Thus, V1 also invites students to collaborate with a colleague to research a term with which we presume they will have no prior familiarity. Again, the research partners are invited to explain what they discover about palimpsests to each other—downloading and printing the example that they like best for eventual use in constructing a poster that will feature a double-bubble chart to compare their downloaded example with the palimpsest they will develop. (A double-bubble chart notionally consists of two intersecting circles as in a Venn diagram— each circle containing a list of characteristics of one of two objects and the intersection containing a list of characteristics that both objects share. Students are often required to list three characteristics in each of the separate and overlapping sections of the circles.) Additionally, students are invited to discuss with their colleague the feelings evoked by the song they have selected and then to decide on symbols that represent those feelings. They are then invited to overlay their symbols—paying attention to balance, variety, unity, and color—on the "canvas" of their chosen song. Table 1.4 is a compilation of the features of the V1 element.

Table 1.3.
Compilation of Salient Features of the M1 Element

Essential Standard 8.ML.2

Interpret the sound and symbol systems of music

Clarifying Objective 8.ML.2.1
Interpret standard musical notation for whole, half, quarter, eighth, sixteenth, and
dotted note and rest durations in 2/4, 3/4, 4/4, 6/8, 3/8, and alla breve meter signatures

Activity Structure (Sandoval, 2014)	Conjectures (Sandoval, 2014)	Computational Thinking (CAS Barefoot)

Task Structure
• Collaborate with a colleague to research the topic "ABC notation" on the Internet and explain your understanding of the essential aspects of ABC notation to each other, taking notes on what you learn.
• When you are satisfied with your understanding, use ABC notation in the EasyABC app (download from https://www.nilsliberg.se/ksp/easyabc/) to play eight bars of a tune of your choice.
• Compose an eight-bar tune of your own either based on the tune you have copied, or by starting anew.
• Create a video Composer's Cut of your MIDI composition project (approx. 3 mins.) in which you
  • discuss your insight into the MIDI environment,
  • play, and
  • evaluate your composition.
 	
Participant Structure
Research partnership
	Design
	If learners engage in M1 using the desktop computers purchased for this purpose and engage in the peer teaching process, then the observable interactions will be conducive to learning and the research partners will each produce both a concordant tune and an acceptable video.
	
	Theoretical
	If the mediating processes materialize, then the research partners will have acquired knowledge commensurate with the expectations of Essential Standard 8.ML.2.
		Concepts
		Research participants will primarily engage in the use of
		• logic (to extrapolate from the research phase of ABC notation to the composition phase)
		• algorithms (to operate within the parameters of Easy ABC
		• evaluation (to make judgements about the MIDI environment and his/her composition
		
		Approaches
		Research participants will individually self-assess the extent of their engagement with the five approaches.

Table 1.4.
Compilation of Salient Features of the V1 Element

Essential Standard 8.V.2
Apply creative and critical thinking skills to artistic expression.

Clarifying Objective 8.V.2.1
Create art that uses the best solutions to identified problems.

Activity Structure (Sandoval, 2014)	Conjectures (Sandoval, 2014)	Computational Thinking (CAS Barefoot)
Task Structure • Collaborate with a colleague to research the topic "palimpsest" on the Internet and explain your understanding of how palimpsests came about. Please download your favorite example. • Use the Internet to find and print the first page of the musical score of one of your favorite songs. • Discuss with your colleague and decide on three different symbols that represent the feeling that your chosen song evokes in you. Layer and interweave your chosen symbols on top of the score to create a palimpsest that is characterized by balance, variety, unity, and color. • Create a poster that features a double-bubble map with three commonalities and three differences for each of the palimpsest you downloaded and your own palimpsest. **Participant Structure** Research partnership	**Design** If learners engage in V1 using the desktop computers purchased for this purpose and engage in the peer teaching process, then the observable interactions will be conducive to learning and the research partners will each produce both an acceptable palimpsest and a poster with a cogent double-bubble map. **Theoretical** If the mediating processes materialize, then the research partners will have acquired knowledge commensurate with the expectations of Essential Standard 8.V.2.	**Concepts** • Research participants will primarily engage in the use of • logic (to generate an artistic design to ensure balance, variety, and unity) • algorithms (to create a layered music score that manifests the stipulated chararacteristics) • patterns (to choose artistically pleasing shapes and colors for their three symbols) • evaluation (to assess their palimpsest in terms of their placement of patterns and color in creating their design) **Approaches** Research participants will individually self-assess the extent of their engagement with the five approaches.

CONCLUSION

Wing (NRC, 2011) criticized what she referred to as "a relatively casual approach" (p. 5) to teaching computer science to middle school students as if they were younger undergraduates. She explained that her pejorative characterization pertained to teaching that was conducted without a deep appreciation of developmental differences across ages. Although she acknowledged a lack of "grounded and research-based knowledge about how the various aspects of computational thinking or computing map on to brain development" (p. 5), she supported what she understood to be a consensus that students need to be in Grade 8 before they can grasp some of the abstract concepts of computational thinking. Wing's perspective was not universally supported by her NRC colleagues, but her perspective on the consensus view lends support for our choice to focus on Grade 8.

Three faculty members attended the parent-teacher organization meetings at two of the middle schools at the end of the first semester to talk about iCS4All and answer parents' questions. A number of parents (including one school board member) and students entered a video-feedback booth after the meeting to voice their deep appreciation for how iCS4All enriched the educational opportunities available at the schools. Much of what they shared recalled for us the references that Denning (2017b) made to the less-than-conscious level at which participants in computational thinking activities engage with the concepts. Not surprisingly, none of the parents who entered the video-feedback booth used the vocabulary of computational thinking to discuss the particular curricular activity element about which they were speaking, but their grasp of the essential aspects of computational thinking was clear—as was their appreciation for the existence of iCS4All and the involvement of the teacher.

In the light of the earlier discussion about the role of education in rural areas, it is vital that we endeavor to be present to the parents to the extent that this is feasible. In addition to attending parent-teacher organization meetings, we specified in our grant submission that we would conduct community learning exchanges to unite the power of place and the wisdom of the people (Guajardo, Guajardo, Janson, & Militello, 2016), the first of which is scheduled as the culminating event of this first full year of our implementation.

In addition to the monthly Moderation Meetings with our teacher partners, we have organized and sponsored the attendance of our school district colleagues at off-site professional development opportunities directly connected with iCA4All. A highlight was a 2-day workshop on design-based implementation research for all the teachers and their principals (mentioned above) which was intended to provide the language and the conceptual space for us to dialogue with our school district partners about how to leverage our already thriving research-practice partnership to sustain iCS4All even beyond the life of this grant. Separately, principals attended a one-day workshop on how to evaluate music and visual art artifacts that they requested be organized for them. In the same vein, teachers attended a 2-day workshop at our university to learn how to operate a 3D printer and acquire the skills to implement a curricular activity system element that was then under discussion. Further, one teacher attended a 5-day workshop during summer that was focused on computational thinking and design thinking in art.

To hearken back to Wing's (NRC, 2011) criticism with which we opened this concluding section, we believe that in iCS4All we have constructed an intentional approach that, with the support of our district, principal, and teacher partners, will continue to enable the Grade 8 students in our three rural school partner districts to employ the pertinent approaches to know and apply the abstract concepts of computational thinking that we have integrated into the music and visual arts standards and generalize their learning to other subject areas.

ACKNOWLEDGMENT

This material is based upon work supported by the National Science Foundation under Grant No. 1738767. Any opinions, findings, and conclusions or recommendations expressed in this material are those of the author(s) and do not necessarily reflect the views of the National Science Foundation.

REFERENCES

Aho, A. V. (2011). Computation and computational thinking. *Ubiquity, 2011, January* (1), 1–8. Retrieved from https://ubiquity.acm.org/article.cfm?id=1922682

Bergeron, P.-J. (2016). How to engage in pseudoscience with real data: A criticism of John Hattie's arguments in *Visible Learning* from the perspective of a statistician. *McGill Journal of Education, 51*(2), 935–945.

Blitstein, R. (2010, February 22). Triumph of the cyborg composer. *Pacific Standard*. Retrieved from http://www.psmag.com/culture-society/triumph-of-the-cyborg-composer-8507/

Carr, P. J., & Kefalas, M. J. (2009). *Hollowing out the middle: The rural brain drain and what it means for America*. Boston, MA: Beacon Press.

Coburn, C. E., & Penuel, W. R. (2016). Research-practice partnerships in education: Outcomes, dynamics, and open questions. *Educational Researcher, 45*(1), 48–54. doi:10.3102/0013189X16631750

Coburn, C. E, Penuel, W. R., & Geil, K. E. (2013). *Research-practice partnerships: A strategy for leveraging research for educational improvement in school districts*. New

York, NY: W.T. Grant Foundation. Retrieved from http://wtgrantfoundation. org/library/uploads/2015/10/Research-Practice-Partnerships-at-the-District-Level.pdf

Computing at School. (n.d.). CAS Barefoot [website]. Retrieved from https:// barefootcas.org.uk/

Corbett, M. (2007). *Learning to leave: The irony of schooling in a coastal community.* Halifax, Canada: Fernwood.

Denning, P. J. (2017a). Computational thinking in science. *American Scientist, 105*(1), 13–17. doi:10.1511/2017.124.13

Denning, P. J. (2017b). Remaining trouble spots with computational thinking: Addressing unresolved questions concerning computational thinking. *Communications of the ACM, 60*(6), 33–39. doi:10.1145/2998438

Duncan, J. K., & Frymier, J. R. (1967). Explorations in the systematic study of curriculum. *Theory into Practice, 6*(4), 180–199. https://doi. org/10.1080/00405846609542084

Eacott, S. (2017). School leadership and the cult of the guru: The neo-Taylorism of Hattie. *School Leadership and Management, 37*(4), 413–426. https://doi.org/10. 1080/13632434.2017.1327428

Fishman, B. J., Penuel, W. R., Allen, A.-R., & Cheng, B. H. (Eds.). (2013). *Design-based implementation research: Theories, methods, and exemplars. National Society for the Study of Education Yearbook.* New York, NY: Teachers College Press.

Guajardo, M. A., Guajardo, F., Janson, C., & Militello, M. (2016). *Reframing community partnerships in education: Uniting the power of place and wisdom of people.* New York, NY: Routledge.

Hattie, J. A. C. (2014). The role of learning strategies in today's classrooms. *The 34th Vernon-Wall Lecture.* British Psychological Society.

Hattie, J. A. C., & Donoghue, G. M. (2016). Learning strategies: A synthesis and conceptual model. *NPJ Science of Learning*, 1–13. Retrieved from https://www. nature.com/articles/npjscilearn201613.pdf

Hattie, J. A., & Donoghue, G. M. (2018). A model of learning: Optimizing the effectiveness of learning strategies. In K. Illeris (Ed.), *Contemporary theories of learning: Learning theorists ... in their own words* (pp. 97–113). London, England: Routledge.

Knochel, A. D., & Patton, R. M. (2015). If art education then critical digital making: Computational thinking and creative code. *Studies in Art Education, 57*(1), 21–38. doi:10.1080/00393541.2015.11666280

Lachman, R. (2018, January 17). STEAM not STEM: Why scientists need arts training. *The Conversation.* Retrieved from http://theconversation.com/steam-not-stem-why-scientists-need-arts-training-89788

McDonald, J., & Allen D. (n.d.) *Tuning protocol: Examining adult work.* Retrieved from https://www.schoolreforminitiative.org/download/tuning-protocol-examining-adult-work/

Militello, M., Jones, K. D., Dunn, L., & Marshburn Moffitt, C. (2018). Learning by leading in the classroom: A rural research-practice partnership. In R. M. Reardon & J. Leonard (Eds.), *Making a positive impact in rural places: Change agency in the context of school-university-community collaboration in education* (pp. 3–32). Charlotte, NC: Information Age.

Mishra, P., & Yadav, A. (2013). Of art and algorithms: Rethinking technology & creativity in the 21st century. *TechTrends, 57*(3), 10–14. doi:10.1007/s11528-013-0655-z

National Academy of Sciences, National Academy of Engineering, & Institute of Medicine. (2007). *Rising above the gathering storm: Energizing and employing America for a brighter economic future.* Washington, DC: The National Academies Press. Retrieved from http://www.nap.edu/catalog.php?record_id=11463

National Research Council. (2010). *Report on a workshop on the scope and nature of computational thinking.* Washington, DC: National Academies Press. doi:10.17226/12840

National Research Council. (2011). *Report on the pedagogical aspects of computational thinking.* Washington, DC: National Academies Press. doi:10.17226/12840

National Science and Technology Council. (2013). *Federal science, technology, engineering, and mathematics (STEM) education strategic plan.* Washington, DC: Author. Retrieved from https://obamawhitehouse.archives.gov/sites/default/files/microsites/ostp/stem_stratplan_2013.pdf

Papert, S. (1980). *Mindstorms: Children, computers, and powerful ideas.* New York, NY: Basic Books.

Peteranetz, M. S., Flanigan, A. E., Shell, D. F., & Leen-Kiat, S. (2017). Computational creativity exercises: An avenue for promoting learning in computer science. *IEEE Transactions on Education, 60*(4), 305–313. doi:10.1109/TE.2017.2705152

Penuel, W. R., Allen, A.-R., Coburn, C. E., & Farrell, C. (2015). Conceptualizing research-practice partnership as joint work at boundaries. *Journal of Education for Students Placed at Risk, 20,* 182–197. doi:10.1080/10824669.2014.988334

Reardon, R. M. (2018). Introduction. In R. M. Reardon & J. Leonard (Eds.), *Innovation and implementation in rural places: School-university-community collaboration in education* (pp. ix–xvii). Charlotte, NC: Information Age.

Robelen, E. W. (2011). Building STEAM: Blending the arts with STEM subjects. *Education Week, 31*(13), 8. Retrieved from https://www.edweek.org/ew/articles/2011/12/01/13steam_ep.h31.html?tkn=UZNF%2BMiYoLmipbRL2XKiLQ0Vmx1S1z7PVU5/&cmp=clp-edweek

Roschelle, J., Knudsen, J., & Hegedus, S. (2010). From new technological infrastructures to curricular activity systems: Advanced designs for teaching and learning. In M. J. Jacobson & P. Reimann (Eds.), *Designs for learning environments of the future* (pp. 233–262). New York, NY: Springer.

Sandoval, W. (2014). Conjecture mapping: An approach to systematic educational design research. *The Journal of the Learning Sciences, 23,* 18–36. doi:10.1080/10508406.2013.778204

Sousa, D. A., & Pilecki, T. (2013). *From STEM to STEAM: Using brain-compatible strategies to integrate the arts.* Thousand Oaks, CA: SAGE.

Smith, M. (2016, January 30). Computer science for all [Blog post]. Retrieved from https://tinyurl.com/y8d5zfpr

Wing, J. M. (2006). Computational thinking. *Communications of the ACM, 49*(30), 33–35. doi:10.1145/1118178.1118215

Wing, J. M. (2016, March 23). Computational thinking, 10 years later. *Microsoft Research Blog*. Retrieved from https://www.microsoft.com/en-us/research/blog/computational-thinking-10-years-later/

Yadav, A., Hong, H., & Stephenson, C. (2016). Computational thinking for all: Pedagogical approaches to embedding 21st century problem solving in K–12 classrooms. *TechTrends, 60*(6), 565–568. doi:10.1007/s11528-016-0087-7

CHAPTER 2

TEACHING A COMPUTER TO SING

Integrating Computing and Music in an After-School Program for Middle School Students

Daniel A. Walzer and Jesse M. Heines
University of Massachusetts

The goal of this work was to research ways in which the teaching of basic computing skills could be integrated into an after-school choral program. The team studied how to adapt the interdisciplinary, computing + music activities developed in an earlier project funded by the National Science Foundation (NSF) with college-aged students to introduce middle school-aged students to computing in an informal, after-school choral program. They investigated how to leverage the universal appeal of music to help students who typically shy away from technical studies to gain a foothold in STEM (science, technology, engineering, and mathematics) by programming

An earlier version of this chapter was published as a paper in the *Journal of Computing Sciences in Colleges, 33*(6), 63–75, June 2018, copyright © 2018 by the Consortium for Computing Sciences in Colleges. The parts of this chapter that appeared in that paper were copied by permission of the Consortium for Computing Sciences in Colleges.

Integrating Digital Technology in Education:
School–University–Community Collaboration, pp. 31–51

choral music. This work was funded by the Advancing Informal STEM Learning (AISL) NSF program, which sought to advance new approaches to, and evidence-based understanding of, the design and development of STEM learning in informal environments. The AISL program included providing multiple pathways for broadening access to and engagement in STEM learning experiences, advancing innovative research on and assessment of STEM learning in informal environments, and developing understandings of deeper learning by participants.

FROM SINGING TO PROGRAMMING

It probably goes without saying that most educators, regardless of field, think it is a good idea for students to be computer literate and even to learn at least a little about how to program. But how do we "hook 'em," especially in an after-school program that does not even begin until after they have been in school for seven or more hours and are ready to just play and hang out with their friends?

Our approach was to work with a dynamic music teacher whom the students adored, having her teach them songs with multiple parts and modest complexity, and then having the students encode those songs in ABC notation (Walshaw, 2017) and program them in Pencil Code (Pencil Code Team, n.d.). (In the first year of our after-school program, we also used Audacity [Audacity Team, 2017] and Scratch [Lifelong Kindergarten Group, MIT Media Lab, n.d.]), but we switched to ABC notation and Pencil Code in our second year. The factors that played into that decision will be discussed subsequently. On occasions, the students "performed" their creations for each other using a computer projector and the room's sound system, thereby both enjoying the sharing of their work and learning from their peers' productions.

While this approach sounds straightforward, it was not easy to implement. Despite their love of listening to—and singing along with—popular music with their friends, the students were initially shy about singing in a class setting. The music teacher worked hard to get them to overcome their reluctance. We who were members of the computer programming team struggled to engage the students in the computer part of the program. Due to our lack of experience with this age group, it took some time to "find our footing." However, persistence paid off and the students indeed made progress in both singing and programming.

This after-school project for middle-school students was the culmination of interdisciplinary research that we began at the college level in 2007. Through the years, 10 university faculty members have been involved in three National Science Foundation (NSF) grants that supported our work, in addition to three middle school teachers and numerous graduate and undergraduate assistants. It truly was a journey of discovery that gave us perspective on what can be accomplished in various settings and enhanced our understanding of the challenges associated with interdisciplinary instruction.

PROJECT BACKGROUND AND GRANT SUPPORT

We were awarded our first National Science Foundation (NSF, 2007) grant to conduct research on the intersection of computing and music in 2007. That project explored interdisciplinary connections among computer science and art, music, and theater (Heines, Jeffers, & Kuhn, 2008; Martin et al., 2009; Ruthmann & Heines, 2009). Given the relationships we were investigating between the performance and informatics fields, we called our work *Performamatics*.

The interdisciplinary collaboration that had the most "traction" for us turned out to be computing+music (Heines, Greher, & Kuhn, 2009; Ruthmann, Greher, & Heines, 2012). That collaboration resulted in a course we called *Sound Thinking* (Greher & Heines, 2009; Greher & Heines, 2014; see the course website at https://jesseheines.com/soundthinking). The popularity and success of *Sound Thinking* led us to apply for our second NSF (2010) grant in this context. That NSF (2010) grant was awarded in 2011, and funded a series of workshops for college teachers on interdisciplinary courses and collaboration (Heines, Greher, Ruthmann, & Reilly, 2011; Heines, Greher, & Ruthmann, 2012; details on these workshops, including all resource materials, are available at https://jesseheines.com/performamatics).

The positive results we observed in both our course and our workshops made us curious about whether we could achieve similar results with younger students. This curiosity led us to apply for our third NSF (2015) grant in this context, which involved working with middle school students. This grant was awarded in 2015 and is the focus of this chapter.

ENABLING RESEARCH AND PROJECT DEVELOPMENT

Underlying Philosophies

John Dewey's (1938) view of the default state of education in his time was one of rigidity in which students were given little (if any) choice in how they were to learn. As he characterized the pervasive approach to education, its "main purpose or objective is to prepare the young for future responsibilities and for success in life.... The attitude of pupils must, upon the whole, be one of docility, receptivity, and obedience" (p. 18). Dewey described the primary aim of education as having students assimilate information, learn new skills, and do so as quietly as possible. According to Maxine Greene (1995), writing 57 years later, Dewey's characterization was still pertinent as

> young people find themselves described as "human resources" rather than as persons who are centers of choice and evaluation. They are, it is suggested, to be molded in the service of technology and the market, no matter who they are. (p. 124)

By contrast, Dewey (1938) championed an experiential approach to learning, while Greene (1995) advocated an approach that inspired and stimulated the imagination, deepened aesthetic experiences, drew inspiration from the arts, and challenged the status quo on rigid definitions of teaching and learning. Both eschewed what they saw as the prevailing understanding that teachers should train students only to be workers and part of a larger economic system. Such an understanding, they asserted, ignored students' socioeconomic status, cultural identity, and personal motivations. Neither Dewey nor Greene opposed vocational training per se, but both advocated for instilling teachers and students with a range of skills drawn from multiple disciplines. At issue is the balance of learning "skills" versus learning critical thinking and problem solving. Like Greene, we believe that a focus on critical thinking and problem solving is more conducive to the development of well-rounded human beings.

The philosophies of Dewey (1938) and Greene (1995) underpinned the development of our *Teach a Computer to Sing* project. One concept to which we kept returning, even as our project experienced a series of setbacks, was that the students *chose* to take part in the project. To maintain that interest, as described subsequently, we transitioned from the teaching model that we used with older students (a laboratory model) to a clubhouse model. Allsup (2016) stated that "at the heart of the laboratory is the impulse to discover and create" (p. 103). The idea of a classroom-as-laboratory inspires a vision of making, doing, and creating. Furthermore, the laboratory is precisely the kind of environment where learning can be messy, revisable, and full of questions. The connections between computing and music did not follow a linear path on all occasions. The same was true of our approach to working with the students. Over the 2-year project our approach evolved considerably, partially because we allowed ourselves to experience the challenge of working with students much younger than those with whom either of us customarily interacted as university professors.

A Rainbow of Partnerships

It is interesting to note that the school with which we worked is named the Bartlett Community *Partnership* School (BCPS, italics added for emphasis). By the time the project ended, we had come to understand "partnership" from multiple viewpoints. As discussed above, this was the result of many years of research among college professors spanning multiple disciplines. As the project got underway, we professors worked closely with the BCPS music teacher, Rachel Crawford, to design an outline of the semester's activities. Rachel's expertise in K–8 music education proved invaluable to us, especially in the first few months. In the second year, our partnership expanded to include Firas Al-Rekabi, a math teacher at BCPS with a PhD in computer science.

We also benefitted from the support of the BCPS principal, Peter Holtz. Peter steadfastly supported our project from the very beginning, advocated for us as we solicited funding, and lent an ear when we needed assistance dealing with logistical issues. At the university level, our site evaluation team from the UMass Lowell Graduate School of Education also provided essential support by observing our work, administering surveys to all participants, and generally providing encouragement when we needed it. The top-to-bottom set of partnerships was essential for us to conduct our research.

Learning Community

We quickly grasped that after a seven or more-hour school day, the students were often tired and hungry. This strongly affected their ability to concentrate during the coding portion of each session. We found that after stating the learning objectives for the day, we needed to allow the students space to form partnerships with their peers and work at their own pace. These partnerships were much easier to establish when we transitioned to a clubhouse environment.

The learning community was further enriched by a third set of partnerships developed between some of our university research assistants and the middle school students. Most of the students were female and persons of color, and some established strong connections with the assistants who shared similar cultural backgrounds.

Our university student assistants also nurtured connections among the university faculty, research team, and the BCPS staff. The student-to-young-adult connection, combined with the balance between singing and coding, established a deeper social bond among all of us. When the students sang, we did too. When they coded and programmed, we sat right next to them and helped. We found it essential that the entire research team "roll up their sleeves" and get involved. This gave us insight into the project's effects and helped us evaluate the students' progress.

We believe that many aspects of the project are best described as "informal learning." Whether the learning community is a "clubhouse" or a "laboratory," a consistent theme emerged in what Allsup (2016) referred to as "teaching for openings," which he described as "the existential turn from obedience to freedom, the aesthetic turn from doing to making, the transformation from past to present, [and] the revelation of self with others" (p. 103). We found that working with the students as partners enabled us to spot entryways to help them connect music and computing through doing. Our interactions with the students helped them build confidence, acquire a modest set of new skills, and heightened our awareness of cultural issues in an urban school setting.

Students, Music, and Computing

Our participants interacted with music and computing in very different ways. We found that some students preferred singing while others preferred coding. Like Dewey (1938), Eisner (1994) argued that there was a "doing" that was required for any "experience" to "stick"—particularly if there was to be some public demonstration or artifact of the learning. We found this to be true not only with the hands-on activities, but also in how we interacted with the students each twice each week (see below for the logistics of the program).

We found it best to find "little moments" of connection—connections between the disciplines and our relationships with the students. As coprincipal investigators, we had to find ways to explain music and computing concepts to students in creative ways. If a student had difficulty hearing harmonies, we might play something on the piano or sing for them. If a student was confused about a coding issue, we often found it helpful to use both music and computing concepts to help them clarify a concept. Many times, the students helped each other work out issues without intervention by one of us or our assistants.

The common thread was that all students had different motivations and ways of experiencing music and computing. There was no singular way to account for how the students perceived the two disciplines. We observed, however, that our pedagogical approach influenced students' motivations to engage with the project. As our twice-weekly interactions with the students evolved, they began to trust us more and more. The most effective "teaching" then happened in small groups and through one-on-one interactions. It was in these personal connections that we found a common language with each learner. We also found that when the students had some ownership over what and how they learned, they were more successful.

PROGRAM LOGISTICS

Teaching a Computer to Sing was an after-school program. It ran for 2½ hours on Tuesdays and Thursdays over the course of 2 school years from

October 2015 through April 2017. Students spent the first half of the session singing songs and then, after a break, spent the second half of the session coding the songs they had sung in the first half of the session using ABC notation (Walshaw, 2017)—developed with the aid of EasyABC (de Jong & Shlien, 2018; see below)—and Pencil Code (Pencil Code Team, n.d.). Sometimes we used the last 10–15 minutes of the computing session for students to show their work to the entire group. At other times, when students were fully engaged in the day's activities, we felt it best to let them continue what they were doing rather than interrupt them to share their creations.

Software Choices

We began the program in 2015 using an earlier version of Scratch (Lifelong Kindergarten Group, MIT Media Lab, n.d.), but the middle school students had considerable trouble converting music notation to Musical Instrument Digital Interface (MIDI) numbers (Wolfe, 2017). This conversion was required because Scratch uses MIDI numbers to play sounds, and, unfortunately, converting musical scores to MIDI involves two steps: (a) figuring out a note's alphabetic representation (such as C), and then (b) converting that alphabetic representation to its MIDI note value (such as 60). Further obscuring the conversion, when it is completed, the resulting Scratch code bears little resemblance to musical notation. A further difficulty is that the resulting Scratch code bears little resemblance to the musical notation. For example, it is hard to know that the code in Figure 1(a) plays "Frère Jacques" even if one compares it to the sheet music in Figure 2.1(b). By contrast, the Pencil Code (Pencil Code Team, n.d.) version in Figure 2.1(c) using ABC notation (Walshaw, 2017)—although it still requires interpretation—is much easier to explain to students than the Scratch version.

To help students make the transition from musical score to ABC notation, we used EasyABC (de Jong & Shlien, 2018). This free downloadable program allows one to enter ABC notation in one window and see the corresponding musical score in a second window. Thus, students can compare the musical score appearing in that second window in EasyABC with the original musical score that they have on paper to ensure that they are the same. If students find that the two musical scores are not identical, the ABC notation is easily edited in EasyABC to correct any problems. There is a learning curve associated with using ABC notation but maintaining the alphabetic representation of the notes is a major advantage. Walshaw (n.d.) explained in the "abc notation blog" that,

> upper case (capital) letters, CDEFGAB, are used to denote the bottom octave (C represents middle C, on the first leger line below the treble [staff],

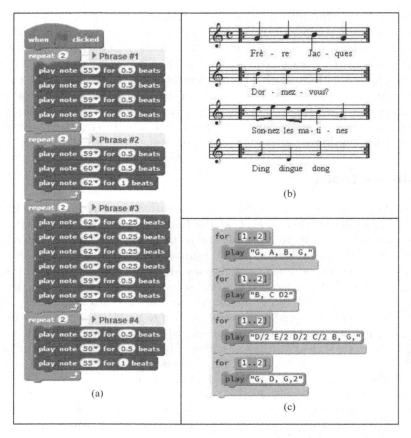

Figure 2.1.   "Frère Jacques" in (a) Scratch, (b) standard music notation, and (c) Pencil Code using ABC notation.

continuing with lower case letters for the top octave, cdefgab (b is the one above the first leger line above the [staff]).

To go down an octave, just put a comma after the letter and to go up an octave use an apostrophe. (How to understand abc (the basics), The notes, paragraphs 3–4)

Figure 2.2 is an annotated example, showing the ABC notation for "Frère Jacques" starting in the bottom octave.

Once the ABC notation (Walshaw, 2017) has been entered into and verified in EasyABC (de Jong & Schlien, 2018), it is easy to copy and paste into Pencil Code (Pencil Code Team, n.d.). The final step is to enclose the Pencil

Code ABC notation in quotes and add code to pass it to a Pencil Code function that plays it. "Pure" Pencil Code using the built-in "play" function is shown in Figure 2.1(c), but a little additional code displayed a dynamic keyboard that showed which key corresponded to the note that was sounded. We wrote a custom function "sing" what allowed playing up to four parts simultaneously by specifying which phrase to play on which piano.

The code in Figure 2.2 (shown in text rather than block mode) calls our custom "sing" function to play "Frère Jacques" as a round (in the key of C) on two keyboards. Note that line 10 rests part 2 for 8 beats (sing 2, "Z8") before "singing" that part, thus creating the round. The lower section of Figure 2.2 also shows the keyboards as they appear during playback.

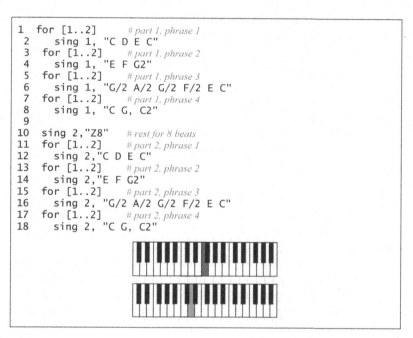

```
1   for [1..2]      # part 1, phrase 1
2     sing 1, "C D E C"
3   for [1..2]      # part 1, phrase 2
4     sing 1, "E F G2"
5   for [1..2]      # part 1, phrase 3
6     sing 1, "G/2 A/2 G/2 F/2 E C"
7   for [1..2]      # part 1, phrase 4
8     sing 1, "C G, C2"
9
10  sing 2,"Z8"     # rest for 8 beats
11  for [1..2]      # part 2, phrase 1
12    sing 2,"C D E C"
13  for [1..2]      # part 2, phrase 2
14    sing 2,"E F G2"
15  for [1..2]      # part 2, phrase 3
16    sing 2, "G/2 A/2 G/2 F/2 E C"
17  for [1..2]      # part 2, phrase 4
18    sing 2, "C G, C2"
```

Figure 2.2.   Playing "Frère Jacques" as a round (starting in the bottom octave).

In the second year, we introduced Soundtrap (2018). This online tool allowed students to record their own songs and sounds to create their own mashups. This is a different type of coding than done with the other tools, but it proved interesting to students and allowed them to extend their work on ABC notation (Walshaw, 2017) and Pencil Code (Pencil Code Team, n.d.). That is, Soundtrap allowed them to capture what they had done in the other programs and remix it along with pre-recorded sounds and effects as well as their own voices to create new compositions.

Song Choices

Songs were mostly chosen by the music teacher. We began with popular songs such Rachel Platten's *Fight Song* (Platten & Bassett, 2013), Taylor Swift's *Shake It Off* (Swift, Martin, & Shellback, 2014), and Shawn Mendes's *Stitches* (Geiger, Parker, & Kyriakides, 2015). We had an arranger create simple harmony parts for these songs, but these did not prove popular because students had trouble with the complexity of the songs' rhythms.

The music teacher suggested that we switch to "partner songs," which were sets of three simple, complementary melodies meant to be sung together. One such set consisted of "One Bottle of Pop," "Don't Throw Your Trash in My Backyard," and "Fish and Chips and Vinegar" (see youtu. be/u-TdsmPHjo0). It was much easier for the students to sing and code these songs in multiple parts.

Student Assistants

One of our project's key features was the relatively large number of university student assistants we employed. We had originally budgeted for two, but it quickly became apparent that that more assistants were required. We found we needed one university student for each pair of middle school students, so we increased the number of assistants to six.

To establish rapport, as mentioned above, the university students (and professors) sang with the middle schoolers in addition to assisting them both with reading music (which was necessary to translate scores into ABC notation) and actual coding. About half the assistants were music majors, while the other half were computer science majors.

In the first year we had only one female assistant: Nicole. With 12 female and two male students in the project, we felt that it was important to employ more female assistants, especially since Nicole provided us with invaluable insights about the complex issues that middle school girls deal with and thus helped us weather a number of storms. In the second year we had three female assistants and two males.

As with any project that employs student assistants, some assistants were more attuned to the context than others. In our case, most student assistants were good at helping to keep the students on task and getting them past hurdles such as computer freeze-ups and simple programming issues. Some made excellent suggestions during our activity planning sessions, and some provided insightful observations when we reviewed each day's experiences.

Some student assistants even established strong personal relationships with the middle school students and functioned as role models, which contributed significantly to the clubhouse atmosphere we were trying to maintain. When one assistant was absent, the students were always disappointed and asked if he or she would be there the next time we met.

Additional Incentives

During the second year, we actively sought ways to motivate the student participants. We produced a CD of their work that proved to be a very popular and motivating activity. Every student contributed at least one track and many contributed multiple tracks. We made about 80 copies to fulfill all of the students' requests, and they gave them to their friends and families for the holidays.

Additional Resources

Throughout the program we created handouts with titles such as "Getting Started with Pencil Code," "Understanding Note and Rest Values," and "Using the TACTS Pencil Code Functions." These handouts, as well as links to demonstration programs and other resources, are available at jheines. github.io/tacts/Workshops, a website we created to support the teacher workshops that we conducted at the 2017 Consortium for Computing Sciences in Colleges (Northeastern Region) and Computer Science Teachers Association conferences.

QUESTIONS AND FINDINGS

Our project investigated two instructional questions and two supporting questions. We attempted to answer these questions both by documenting our own informal observations and by employing surveys and focus groups conducted by the evaluation team. Our own informal observations obviously suffered somewhat from observer bias, while the survey and focus groups suffered somewhat from small sample size. Nevertheless, we believe that we can draw a number of conclusions from our experiences that can be sustained by credible anecdotal evidence if not by hard numbers and statistical significance.

Instructional Questions

Question 1. Can middle school students follow the connections from singing to digitized sound to computer notation and back to music to help them learn to program songs they like to sing?

Our answer to this question is an unqualified "yes." We would go even further to say "yes, and sometimes with enthusiasm." Of course, not every student was "into" the coding part of the project, and some days it seemed like no one wanted to do any coding at all. (As every parent knows, such is life when working with children.) However, on other days, a good number of students exhibited real excitement about their ability to "teach a computer to sing" and eagerly lined up for "show and tell" at the end of the day to demonstrate their accomplishments to others.

As stated above, we began the year having students code music as a series of Pencil Code (Pencil Code Team, n.d.) play blocks. As the year progressed, we introduced progressively more advanced computing concepts. The first of these was simple loops (like those in Figure 2.2(c)), which allowed students to repeat musical phrases. Next, we introduced loops with control variables, such as "for k in [1..3]." This allowed students to make the connection between songs with larger repetitive structures that used first and second endings by coding conditional "if" statements. We then introduced general variables, which allowed us to store and reuse musical phrases coded as ABC notation strings. A couple of students even got as far as indexed variables (one-dimensional arrays), which we used to code two parallel structures in order to pair each note with its lyric.

Only one student in our project was in Grade 7, while all the other students were in either Grade 5 or Grade 6. Only one student had prior experience with computer programming, and about 25% did not have computers at home. Given these demographics, the fact that the students had already been in school for seven or more hours by the time they began programming with us, and our own inexperience in teaching middle school students, we feel that the programming concepts we were able to introduce represent a reasonable level of learning.

Question 2. Can students learn to sing three- and four-part harmony by programming the individual parts?

During initial discussions, the music teacher told the professors that she thought it would be difficult for the students to sing in more than two parts because they had never done it before. By the end of the year, however, they were successfully singing the "partner songs" in three parts. They even cheered when they all finished at the same time! We cannot claim all the credit for that progress, but the music teacher attributed at least some credit to our project.

One of the programming techniques that seemed to help students learn multipart songs was the use of variables to store and reuse musical phrases coded as ABC notation strings. This helped students see song structures, notice where phrases repeated, and understand how the melody lines went together. Again, we do not want to overstate this result, as it was impossible to measure objectively. However, we have learned to rely on the music teacher's perception in this regard, as she is highly attuned to the students' capabilities.

Supporting Questions

Question 3. What resources, models, and tools are necessary to integrate STEM into a middle school, after-school choral program?

The resources initially available at BCPS were severely limited. The computer network and Internet access were so severely configured that Windows systems could not access the network at all, and sites such as YouTube were not accessible without teacher credentials. Luckily, access to all the music sites we utilized was supported.

We were also unable to install software on the school systems. No one in the school had authority to do so, either. To install software, we would have had to make a request of the central school district office, and wait weeks of it to be addressed. To resolve this issue, we were fortunate in being able to buy systems specifically for our project's exclusive use.

As noted earlier, our model for the after-school project was initially what we were used to: a laboratory class. As mentioned above, this did not prove viable, as the students were simply unable to pay attention to instructions for more than a minute or two in the after-school environment. We therefore transitioned toward a clubhouse model, where students worked one-on-one or two-on-one with a professor, university student assistant, or another middle school student.

Also as noted earlier, we prepared handouts with instructions and illustrations so that the students could work on their own rather than listen to us explain how to accomplish the day's goals. We also hired three times the number of university student assistants that we had originally planned, as it became evident that they were needed in the clubhouse model.

The tools we used have been discussed previously, but it is important to reemphasize that they changed throughout the program. The switch from Scratch (Lifelong Kindergarten Group, MIT Media Lab, n.d.) to Pencil Code (Pencil Code Team, n.d.) was the biggest unanticipated change at the beginning of the first year, and the discovery of EasyABC (de Jong & Shlien, 2018) for writing ABC notation proved to be a godsend.

Question 4. Can the involvement of student assistants and teachers who match the students' racial and/or cultural backgrounds have a positive effect on a putative sense among some students that "people like me don't (or can't) do that"?

One of the issues that concerned us was our ability to "connect" with the middle school students. Almost all of the students had very different racial profiles from our own, and, as noted earlier, the vast majority (86%) were female. Numerous authors such as Kohl (1994), Delpit (2006), and Tatum (1997) have written about issues related to race in the classroom, and we feared that at least some of those issues extended to gender, as well.

Looking back, it appears that we need not have been so concerned, as racial and cultural attributes did not appear to be major stumbling blocks. The students built relationships faster with some university students than with others, or with the professors, but, over time, all facilitators were able to build relationships with all of the students.

Peer pressure seemed to play a larger role in relationship-building. In follow-up interviews conducted by our evaluation team, one professor observed that he had no trouble working with any student one-on-one. As soon as two or three of the students got together, however, they seemed to shut him out. He said he felt that the students—especially the girls—appeared reluctant to behave in such a way that their peers might conclude that they were comfortable talking to a teacher, especially an older White male.

The bottom-line answer to this question is that, while we can only report observational and anecdotal evidence, in our case we believe that the premise of this question does not appear to be supported. That said, we refrain from generalizing since our project had a small sample size and that our results are dependent on the many specific personalities involved.

LESSONS LEARNED

Based on these experiences, the main lessons we learned are as follows.

- Working with students after they have been in school for seven or more hours is hard. There are times when one has to just let them play. In addition, one must understand that some days students just will not want to code, and, on such days, pushing them to do so is futile. There were even days when the beloved and experienced music teacher found it difficult to get them to sing. "Go with the flow."
- Knowing how students perceive a project such as this is also hard. Attitudes often cannot be seen, and one must be very careful not to make assumptions based on observed behaviors.
- When meeting only once or twice a week, there is a strong need for concrete, over-arching goals to tie sessions together.
- University student assistants must be vetted carefully. We had no major problems with any assistants' interactions with the middle school students, but it was clear that some were far better than others at helping the middle school students learn and remain on task.

- There is simply no substitute for partnering with an experienced teacher who has a strong rapport with the students and understands where they are coming from. On several occasions, our music teacher partner pointed out where some of our assessments and impressions of how the project was going were wrong. For example, we thought that one student who did not seem to engage with the project simply did not want to be in it. Our teacher partner pointed out that all this student's friends had dropped out of the project for one reason or another, and the fact that she was still with us was a strong indicator of her desire to be there.
- Despite all the time and patience, it takes to get students to focus on learning in an after-school program and the inevitable ups and downs of such an endeavor, there are numerous, priceless, unforgettable "ah-ha" moments that make the effort worthwhile.

FORMAL EVALUATION

Our project was supported by an evaluation team centered in the UMass Lowell Center for Program Evaluation. The team observed after-school sessions, administered surveys, and conducted focus groups with both students and faculty. We present just a few of their more interesting findings here in descriptive terms, because all but one of the statistical measures were not significant due to small sample sizes.

Student Surveys and Focus Groups

Student surveys and focus groups revealed two major outcomes, although we stress again that these are not statistically significant.

- Measures of students' attitudes toward music and their perceptions of their own music-related abilities both increased slightly from pre- to post-program assessments.
- Measures of students' attitudes toward computer programming remained the same, but their perceptions of their own computer-related abilities increased slightly from pre- to post-program assessments.

To demonstrate just how tricky it is to conduct formal evaluation on this type of program, consider the following student responses to open-ended questions. The reasons cited most frequently for liking computer programming were "making games, music, and websites," "coding," and "I don't know." Those cited most frequently for not liking computer programming were "boring," "hard," "so much to do," and "I don't know." We find it interesting that "coding" shows up in the positive list, while "hard" and "so much to do" show up in the negative list. These seem contradictory, and of course the "I don't know" response is not helpful, particularly because it appears in both lists.

A more positive outcome that the evaluation team reported was that 67% of students responded "yes" when asked if computer science and music were related in any way. We interpret this as at least an indication that most students were able to achieve one of our primary goals: to have them follow the connections from singing to digitized sound to computer notation and back to music.

The focus group discussion clearly revealed that students preferred working with EasyABC (de Jong & Shlien, 2018) and Pencil Code (2016) to Audacity (Audacity Team, 2017) and Scratch (Lifelong Kindergarten Group, MIT Media Lab, n.d.). They liked the visual aspects of EasyABC and the versatility of Pencil Code, that is, its ability to support various types of tasks.

It is also telling that the students' two most prominent suggestions across both years for improving the program were related to having more snacks and more fun. We conjecture that such comments might be typical for an after-school program. We consciously accommodated the students' appetites by allowing snacks to be eaten during the project activities and giving students more free time after they completed coding activities.

Facilitator Surveys and Focus Groups

As discussed above, a number of changes were made in Year 2 of the program. Facilitator surveys and focus groups were conducted to assess the effect of these differences, at least from the facilitators' perspective.

Using a 5-point Likert scale, facilitators were asked whether students were able to follow the connections from singing to digitized sound to computer notation and back to music. As shown in Table 2.1, their responses were far more positive in Year 2 than in Year 1, and even with the small sample size the difference was statistically significant ($p < .05$).

Facilitators were also asked if they felt that (a) using songs that students like to sing helped them learn to program, (b) students' affect toward programming had improved throughout the program, and (c) working with adults whose racial or cultural backgrounds matched the those of the students had a positive effect on the students. The Likert scale results for these questions increased from Year 1 to Year 2, as shown in Table 1, but not enough to be statistically significant.

Table 2.1.
Facilitators' Perspectives on Project Objectives

Objective	Year	N	Mean	Std. Error	Sig.
Follow connections "both ways"	1	4	2.00	0.577	$p < .05$
	2	8	3.88	0.350	
(a) Learn to program using songs	1	4	3.50	0.289	no
	2	8	4.13	0.295	
(b) Improve affect towards programming	1	4	3.50	0.289	no
	2	7	3.71	0.184	
(c) Racial/cultural matching positive	1	4	2.75	0.479	no
	2	7	3.71	0.184	

The differences between facilitators' responses to open-ended questions in Year 1 and Year 2 are telling with regard to the project's development.

- In Year 1, responses focused on what they needed to teach. In Year 2, they focused on the need for more structured plans.
- In Year 1, their favorite songs to work on were the partner songs. In Year 2, their favorites were those that supported loops.

A major theme in the focus group discussion was relationships, that is, establishing a rapport with the students so that they would be receptive to instruction. This improved greatly in Year 2 with more careful vetting of university student assistants and the professors' acceptance of the need to work one-on-one or one-on-two. In addition, we all had to learn to allow for the ebb and flow of the students' learning and attention span in an after-school program.

CHALLENGES FOR COLLEGE PROFESSORS

One aspect of our work that we have not seen discussed in the litera-ture involves how interdisciplinary collaborations are viewed by different college departments, particularly with regard to promotion and tenure (P&T) decisions. In science and engineering departments, virtually all work is collaborative. Thus, coauthored papers typically carry the same "weight" in P&T decisions as single-authored papers, particularly if one is listed as the first, second, or sometimes even the third author. In some arts departments, however, the only work that "counts" may be work that one has done solely by oneself, such as a performance, show, piece of art, or composition. The undervaluing of collaborative projects discourages interdisciplinary work, as it can actually diminish a participant's standing with respect to P&T in some arts departments.

On the other hand, a benefit for arts professors in partnering with science and engineering professors is the dollar amount of the grants available. A $100,000 research grant in the sciences is typically considered a small grant, while such a grant would be considered huge in most arts departments. Larger grants typically provide more flexibility in how one conducts research, thus allowing arts professors to be freer and more creative in pursuit of their goals. The bottom line here is that we encourage all professors who aspire to promotion and tenure to check with their departments before following in our footsteps.

A second consideration is the time required to work with students in an after-school program. As mentioned earlier, we were at BCPS for 3 hours on each of two afternoons per week, and the times that we needed to be there were immutable. We found this to be a demanding schedule, even though BCPS was only a mile from our offices. Such a commitment may or may not be a problem, depending upon one's teaching and meeting schedule and other university obligations. In keeping with our comment regarding promotion and tenure, however, we recommend that professors check with their departments before making such a commitment of their time.

RECOMMENDATIONS FOR FUTURE PROGRAMS

- Structure the program to include ample time for facilitator planning and preparation outside of the time spent with the student participants.

Comment: This was difficult for us given our office locations of different campuses and the differences in our teaching schedules. We also would have benefitted from more outside the classroom time with the university student facilitators.

- Have many structured activities available to keep the students engaged.

Comment: In most sessions, we had structured activities for the main theme of the day, but it would have been good to have had several backup structured activities, as well.

- Ensure that all facilitators circulate throughout the room and work with the students one-on-one or in two-on-one.

Comment: Some of the university student assistants thought that they should wait until a student asked for assistance, which the students seldom did. We talked with the student assistants about this, but some were still reluctant to simply sit down next to a participating student and ask him or her to share what he or she was doing. On the other hand, the participating students were very receptive to help and advice when it was offered.

- Present short narratives or movie clips and discuss famous individuals in STEM fields who were from minority backgrounds.

Comment: The students were very receptive to the clips that we did show them, and it would have been good to have done that more often.

- Provide many opportunities for students to share what they have done in response to the project tasks.

Comment: The students were all strongly engaged with the creation of the holiday CD. In retrospect, we could have created a second CD or had some other culminating joint project to further motivate their engagement.

ACKNOWLEDGMENT

This chapter is based upon work supported by the National Science Foundation under Grant No. 1515767. from the National Science Foundation Division of Research on Learning. Any opinions, findings, and conclusions or recommendations expressed in this chapter are those of the authors and do not necessarily reflect the views of the National Science Foundation.

Earlier work referenced in this chapter was supported by National Science Foundation under Grant No. 0722161 and Grant No. 1118435. Gena Greher of UMass Lowell and S. Alex Ruthmann of NYU contributed substantially to our background work.

We are deeply indebted to Rachel Crawford, Firas Al-Rekabi, and Peter Holtz of the Bartlett Community Partnership School for their invaluable contributions to this project. Without a doubt, it could not have been done without them.

REFERENCES

Allsup, R. (2016). *Remixing the classroom: Toward an open philosophy of music education*. Bloomington IN: Indiana University Press.

Audacity Team. (2017). Audacity®: Free audio editor and recorder [Computer application]. Retrieved from https://www.audacityteam.org

Delpit, L. (2006). *Other people's children: Cultural conflict in the classroom*. New York, NY: The New Press.

Dewey, J. (1938). *Experience & education*. New York, NY: Touchstone.

de Jong, J. W., & Shlien, S. (2018). EasyABC [Computer application]. Retrieved from https://sourceforge.net/projects/easyabc/

Eisner, E. W. (1994). *Cognition and curriculum reconsidered*. New York, NY: Teachers College Press.

Geiger, T., Parker, D., & Kyriakides, D. (2015). *Stitches* [Recorded by Shawn Mendes]. Wisconsin: Hal Leonard Music.

Green, M. (1995). *Releasing the imagination: Essays on education, the arts, and social change*. New York: NY: Jossey-Bass.

Greher, G. R., & Heines, J. M. (2009, October). *Sound thinking: Conceptualizing the art and science of digital audio for an interdisciplinary general education course*. Paper presented at the conference of the Association for Technology in Music Instruction (ATMI), Portland, OR.

Greher, G. R., & Heines, J. M. (2014). *Computational thinking in sound: Teaching the art & science of music & technology*. New York, NY: Oxford University Press.

Heines, J. M., Jeffers, J., & Kuhn, S. (2008). Performamatics: Experiences with connecting a computer science course to a design arts course. *International Journal of Learning, 15*(2), 9–16.

Heines, J. M., Greher, G. R., & Kuhn, S. (2009). Music performamatics: Interdisciplinary interaction. In *Proceedings of the 40th Association for Computing Machinery (ACM) Technology Symposium on Computer Science Education* (pp. 478–482). Chattanooga, TN: ACM.

Heines, J. M., Greher, G. R., Ruthmann, S. A., & Reilly, B. (2011). Two approaches to interdisciplinary computing+music courses. *IEEE Computer, 44*(12), 25–32.

Heines, J. M., Greher, G. R., & Ruthmann, S. A. (2012, July). *Techniques at the intersection of computing and music*. Paper presented at the 17th Annual Conference on Innovation & Technology in Computer Science Education (ITiCSE). Haifa, Israel.

Kohl, H. (1994). *"I won't learn from you" and other thoughts on creative maladjustment*. New York, NY: The New Press.

Lifelong Kindergarten Group, MIT Media Lab. (n.d.). Scratch [Computer application]. Retrieved from scratch.mit.edu

Martin, F., Greher, G. R., Heines, J. M., Jeffers, J., Kim, H.-J., Kuhn, S., ...Yanco, H. (2009). Joining computing and the arts at a mid-size university. *Journal of Computing Sciences in Colleges, 24*(6), 87–94.

National Science Foundation. (2007). Performamatics: Connecting Computer Science to the Performing, Fine, and Design Arts.   Award #0722161, 2007-2010, $421,087.   http://www.nsf.gov/awardsearch/showAward?AWD_ID=0722161.

National Science Foundation. (2010). Computational Thinking Through Computing And Music. Award #1118435, 2011-2015, $499,995. http://www.nsf.gov/awardsearch/showAward?AWD_ID=111843.

National Science Foundation. (2015). A Middle School After-School Pilot Program Integrating Computer Programming and Music Education. Award #1515767, 2015-2017, $288,945. http://www.nsf.gov/awardsearch/showAward?AWD_ID=1515767.

Pencil Code Team. (n.d.). Pencil Code [Computer application]. Retrieved from pencilcode.net

Platten, R., & Bassett, D. (2013). *Fight Song* [Recorded by Rachel Platten]. Platten Music.

Ruthmann, S. A., & Heines, J. M. (2009, October). *Designing music composing software with and for middle school students: A collaborative project among senior computer science and music education majors.* Paper presented at the conference of the Association for Technology in Music Instruction (ATMI), Portland, OR.

Ruthmann, S. A., Greher, G. R., & Heines, J. M. (2012, July). *Real world projects for developing musical and computational thinking.* Paper presented at the 30th World Conference of the International Society for Music Education (ISME) on Music Education. Thessaloniki, Greece.

Soundtrap. (2018). Soundtrap education [Computer application]. Retrieved from www.soundtrap.com/edu/

Swift, T., Martin, M., & Shellback. (2014). *Shake it off* [Recorded by Taylor Swift]. Nashville, TN: Sony/ATV Music.

Tatum, B. D. (1997). *"Why are all the black kids sitting together in the cafeteria" and other conversations about race.* New York, NY: Basic Books.

Walshaw, C. (2017). *abc notation.* Retrieved from abcnotation.com

Walshaw, C. (n.d.). abc notation blog: How to understand abc (the basics) [Blog post]. Retrieved from http://abcnotation.com/blog/2010/01/31/how-to-understand-abc-the-basics/

Wolfe, J. (2017). *Note names, MIDI numbers and frequencies.* Retrieved from newt.phys.unsw.edu.au/jw/notes.html

PART II

THE DIGITAL TECHNOLOGY EDUCATOR

CHAPTER 3

A MULTIDISCIPLINARY APPROACH TO INCORPORATING COMPUTATIONAL THINKING IN STEM COURSES FOR PRESERVICE TEACHERS

Jennifer E. Slate, Rachel F. Adler, Joseph E. Hibdon, Scott T. Mayle, Hanna Kim, and Sudha Srinivas
Northeastern Illinois University

Despite being part of the science and engineering practices defined by Next Generation Science Standards (NGSS) guidelines, computational thinking is rarely addressed in preservice teacher training. Motivated by this gap, we modified our teacher education program to embed computational thinking throughout the curriculum for preservice elementary and middle school teachers. We integrated computational thinking into five courses, including (1) an introductory computer science course added to the program, (2) redesigned content courses in biology, physics, and geometry with newly created computational thinking modules, and (3) a modified science teaching methods course in which preservice teachers learn to integrate computational thinking into their future classrooms. This three-tiered approach,

Integrating Digital Technology in Education:
School–University–Community Collaboration, pp. 55–80
Copyright © 2019 by Information Age Publishing

which presents computational thinking within the context of discipline-specific content and pedagogical technique, was achieved by close collaboration among education, computer science, math, and science faculty. In pre- and post-surveys, preservice teachers reported significant gains in their perceived knowledge and ability in computational thinking. Interest and confidence in incorporating computing and computational thinking into future teaching was high at both the beginning and end of the study. These preservice teachers are now better positioned to foster computational thinking skills and practices among children in their future classrooms.

The Next General Science Standards (NGSS) call for a radical shift in K–12 science education (National Research Council [NRC], 2012). Textbooks and lessons too often focus on definitions or rote problem-solving, and so students perceive STEM disciplines as consisting of unrelated facts to memorize and formulas to calculate. NGSS recognizes that children are natural investigators and advocates, and that they develop questions, test ideas, and evaluate evidence. In addition to identifying core ideas for STEM disciplines and cross-cutting concepts that connect the disciplines, NGSS recommends practices that engage children in inquiry.

Computational thinking plays a key role in the "NGSS Science and Engineering Practices" (n.d.), one of which is "using mathematics and computational thinking" (para. 6). While the practice of mathematics is well established, the concept of computational thinking may be unfamiliar to many teachers (Yadav, Stephenson, & Hong, 2017). This deficit must be remedied, because computer technologies have revolutionized not only science and engineering, but society as a whole.

Although today's children are often called the app generation due to their savvy with tablets and smartphones, using those tools is not the same as understanding computational thinking (Leonard et al., 2016). An analogy given by Wing (2008) is that using a calculator is not the same as understanding arithmetic. To harness the technological power of computers to solve problems, future generations will need to be able to frame problems in ways that allow computers to be used to explore solutions.

The development of computational thinking skills in children will require teacher training. Yadav, Stephenson, and Hong (2017) recommend that this training begin with preservice teachers and that education and computer science faculty work together to revise teacher development programs. Accordingly, as an interdisciplinary team, we embedded computational thinking into curricula for future elementary and middle school teachers. To redesign our teacher training program, we (1) created an introductory computer science course, (2) integrated modules into biology, physics, and geometry courses that apply computational thinking to content, and (3) modified a science teaching methods course so that preservice teachers learn to incorporate computational thinking into their future classrooms.

COMPUTATIONAL THINKING

What Is Computational Thinking?

Computational thinking is an analytical approach to thinking that draws upon computer science concepts. The International Society for Technology in Education (ISTE) and the Computer Science Teachers Association (CSTA) developed the following description that educators can use to incorporate computational thinking into content areas across the curriculum (ISTE & CSTA, 2011):

- Formulating problems in a way that enables us to use a computer and other tools to help solve them,
- Logically organizing and analyzing data,
- Representing data through abstractions such as models and simulations,
- Automating solutions through algorithmic thinking (a series of ordered steps),
- Identifying, analyzing, and implementing possible solutions with the goal of achieving the most efficient and effective combination of steps and resources, and
- Generalizing and transferring this problem-solving process to a wide variety of problems. (para. 2)

Computational thinking should not be confused with mathematical computation or with the simple use of computers. It is not the plugging of numbers into a formula to calculate an answer, nor is it just the use of technological tools. For example, simply using the Internet to look up information is not computational thinking. However, computational thinking skills are required to break down a problem into its parts, identify key components, and to logically organize and analyze relevant information obtained about those components.

Common teaching tools in K–12 education can be used to help build computational thinking skills (Mouza, Yang, Pan, Ozden, & Pollock, 2017). For example, interactive whiteboards allow children to engage with a problem and help formulate it in a way for it to be solved. Graphical organizers, used to visualize complex concepts or systems, represent conceptual models. In this instance, children may draw a concept map to illustrate relationships among plants and animals in a food web. Finding common characteristics that allow placement of the organisms into groups requires abstraction. Putting those groups in proper order (e.g., herbivores eat plants, and predators eat herbivores) requires algorithmic thinking.

In these ways, children can begin to build computational thinking skills without necessarily learning to write computer code (Barr, Harrison, & Conery, 2011). Before programmers begin to code they must first decompose a problem into parts and conceive of potential steps for solving the problem. Computational thinking should not be thought of as synonymous with programming, but as a problem-solving process that precedes the writing of code (Wing, 2008).

Although computational thinking involves more than programming, learning to write a computer program can further develop computational thinking skills (Voogt, Fisser, Good, Mishra, & Yadav, 2015). Automating the solving of a problem requires creating computer code in a series of ordered steps, which is an algorithm. An efficient and effective combination of steps can be determined by implementing and analyzing various coding solutions. Finally, code written to solve one problem can be generalized and transferred to another problem. For example, a computer simulation of the transfer of energy between an animal and its food source could be modified to model the transfer of energy between a car and its fuel.

The thinking process involved in formulating a problem so that it can be solved by a computer is a skill relevant across disciplines. Because computers are pervasive in nearly all fields, children should learn how computers can be used to solve problems—when pursuit of a computer-based solution is warranted—and how to communicate with those programming the computer technologies (Barr et al., 2011). To ensure success in the modern workplace, computational thinking should become a part of every child's analytical ability and be as important in the K–12 curriculum as reading, writing, and arithmetic (Wing, 2006).

Computational Thinking in Elementary and Middle School Education

The Computer Science for All movement calls for all K–12 students to engage in high-quality computer science learning (CSforALL.org). Teaching young children to think computationally, beginning in elementary and middle school, will help bring about this goal. In addition to preparing students for learning computer science, the problem-solving approach of computational thinking illustrates its relevance. Grover, Pea, and Cooper (2014) found that most middle school students thought that computer science was tantamount to the building or fixing of computers.

After viewing examples of computer scientists solving real-world problems in creative ways, the children came to understand that computers are tools that can be programmed to improve people's lives.

Computational thinking can be introduced during in-school and after-school contexts, through robotics, game design, and modeling and simulation (Lee et al., 2011). Constructible robotics kits with drag-and-drop programming blocks allow children as young as kindergarten to design and program robots. They think algorithmically to place coding blocks in sequence and learn to identify, analyze, and implement possible solutions while debugging (Bers, Flannery, Kazakoff, & Sullivan, 2014). Scratch, another introductory programming platform with a drag-and-drop inter-face, quickly became the most popular activity choice in an after-school technology center for urban youth (Maloney, Peppler, Kafai, Resnick, & Rusk, 2008). This was despite the availability of alternatives such as video-games and image manipulation software. Instead, children used Scratch to create their own games and images.

Problem-solving skills are strengthened through computational thinking activities. To program a moving object, children have to make decisions about angles, length measurements, and x- and y-coordinates, just as they must do when solving a problem in geometry (Calder, 2010). To design a robot and predict how far it will travel once it is programmed, children must compare distance traveled per rotation for wheels of different sizes (Grubbs, 2013).

Modeling and simulation not only build computational thinking skills, but promote scientific inquiry (Sengupta, Kinnebrew, Bau, Biswas, & Clark, 2013). Computer models and simulations allow children to conduct experiments repeatedly and immediately visualize results. They can examine topics that are unrealistic to practically test, such as the effect of mass on planetary orbits (Ornek, 2008). Computer simulations also help students grasp complex systems. Building a model of a fish tank required that middle school students identify key components and program each with appropriate behaviors (e.g., fish produce waste and plants consume waste). When the model was not sustainable, the children realized the need to include bacteria, which convert the fish waste into a form usable by plants (Sengupta et al., 2013).

Computational thinking also enhances learning outside of STEM. Being able to develop a logical argument by abstracting information from known cases and putting that information in order is a desired skill in a variety of subjects. Middle school students used such computational thinking skills to write journalism stories and social studies reports, which they enhanced by programming animations with Scratch (Wolz, Stone, Pearson, Pulimood, & Switzer, 2011). They also applied computational thinking to write and program choose-your-own adventure stories (Settle et al., 2012). Such stories are nonlinear, requiring students to abstract the plot by finding commonalities in different plot threads so that pieces of the story can be reused. Children also learned to think computationally while programming an avatar to dance along with them in a performance (Leonard & Daily, 2013). To choreograph the dance, they abstracted complex moves into key steps and organized the steps into units to be put together.

A Use-Modify-Create approach can ease children into the process of computational thinking (Lee et al., 2011). They begin by using a pre-existing robot, game, or computer model. After becoming familiar with the code and how it works, they can begin to make modifications. Modifications may at first be simple (e.g., changing a character's color), but eventually more complex (e.g., changing a character's behavior). Finally, students gain the skills and confidence to create new computational projects about their own ideas.

We are far from being able to provide Computer Science for All, despite widespread interest among parents and children. Only 40% of K–12 schools offer computer science, even though 82% of students have at least some interest and 84% of parents believe it is as important as required subjects like math, science, history, and English (Wang, 2017). A lack of qualified teachers is the most commonly cited barrier reported by principals and superintendents. All teachers will benefit from becoming proficient in computational thinking, not only to provide children a foundation in computer science, but to engage them in problem-solving skills that are applicable across disciplines. Unfortunately, few teacher education programs currently include computational thinking, and so preservice teachers are not trained to apply it in their future classrooms (Yadav et al., 2017).

REVISED PRESERVICE EDUCATION CURRICULUM

Our modifications to the curriculum for preservice teachers were built upon the Technological Pedagogical and Content Knowledge (TPACK) framework (Koehler, Mishra, Kereluik, Shin, & Graham, 2014). TPACK recognizes the role of technology in education and its intersection with pedagogical and content knowledge. Understanding and using technological platforms can improve teachers' pedagogical practices and students' learning of content. Thus, technology should not be taught to preservice teachers in isolation, but in relation to how it can be applied in teaching (Mishra & Koehler, 2006).

Yadav et al. (2017) discuss how TPACK can guide the integration of technology into teacher education programs. Preservice teachers should build the ability to think computationally, gain an understanding of computational thinking within the context of subject areas, and learn to integrate computational thinking into their future classrooms. Teacher training programs can incorporate computational thinking by modifying existing curricula in courses such as educational technology or teaching methods, or by creating stand-alone courses that involve computational thinking concepts and tools. Partnerships between education and computer science faculty are key during this process, to expose both educators and students to concepts and practices used by computer scientists and to develop appropriate learning outcomes.

Our teacher education program is at a public, urban, and highly diverse Midwestern university. We are a federally designated Hispanic serving institution with an undergraduate student population that is about 37% Hispanic, 28% Caucasian, 12% African American, 8% Asian, and 0.2% Native American (other students self-reported as multiracial or did not indicate race or ethnicity). Low-income students comprise approximately 43% of undergraduates, and about 44% of all students are the first in their families to attend college. Preservice teachers enrolled in local community colleges also take courses in our program alongside the university students and will ultimately transfer through a bridge program.

We began the integration of computational thinking into the curriculum with an activity developed for science teaching methods (Adler & Kim, 2018) and subsequently expanded our efforts to include additional courses for preservice teachers. Working together as a team of education, computer science, math, and science faculty, we developed a three-tiered approach (see Figure 3.1). We created a foundational computer science course with a focus on applications in education, developed modules to build computational thinking skills within the context of STEM content courses, and further modified the science teaching methods course to show preservice teachers how to apply computational thinking to their future classrooms. All of the courses were taken before preservice teachers conduct teaching practicums.

CS0 Computer Science Course for Educators

To provide preservice teachers with a foundation in computational thinking and programming, we developed an introductory computer science CS0 course. CS0 courses are increasingly used to introduce computer science concepts to students who are not ready to take Computer Science I (Anewalt, 2008; Powers, Ecott, & Hirshfield, 2007; Uludag, Karakus, & Turner, 2011). CS0 courses can be designed to prepare students for a computer science major (Rizvi, Humphries, Major, Jones, & Lauzun, 2011) or to interest nonmajors in programming concepts and applications (Cliburn, 2006).

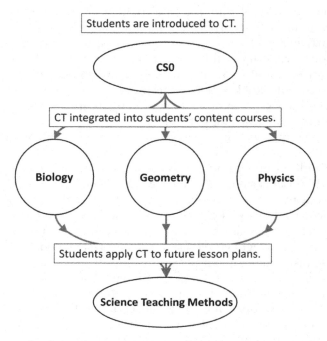

Figure 3.1.   Three-tiered approach for the inclusion of computational thinking (CT) into teacher training, beginning with a foundational computer science (CS0) course.

We began by teaching preservice teachers to think computationally (Cortina, 2007). Through everyday examples, they learned to "think" like a computer. To create an algorithm, they wrote in proper order the steps necessary to brush teeth. Leaving out a crucial step, such as removing the cap from the toothpaste tube, would prevent a robot from being able to perform the task. To further formulate problems in a way that would enable a computer to solve them, preservice teachers built flowcharts. Additional skills such as abstraction and the finding of patterns in data were practiced with activities from Google's Computational Thinking for Educators course (https://computationalthinkingcourse.withgoogle.com). For example, in a fun variation of the game 20 questions, they figured out how many guesses it would take to identify any species on Earth.

Preservice teachers were then introduced to programming through Hour of Code (https://studio.code.org/courses), a worldwide effort to demystify the writing of computer code and show that anybody can learn coding basics. Coding techniques such as creating loops, conditionals, and variables were practiced through visual-based programming exercises that did not require the learning of technical terms.

To further develop coding skills, preservice teachers created programs with three platforms transferable to K–8 classrooms: Lego Mindstorm robots, Scratch, and VPython. Lego robots snap together to allow children to create new ideas and structures (Bers et al., 2014). The Scratch platform was designed to have a similar feel to Lego robots, as children snap together graphical programming blocks on the screen to create programs (Resnick et al., 2009). Both use drop-and-drag commands and thus do not require the learning of coding language. So that preservice teachers would not later fear or shy away from text-based code, we also introduced VPython. VPython is easier to learn than many programming languages due to its intuitive commands. It also has a 3D graphics interface that allows visualization of scientific models (Ornek, 2008). Scratch and VPython are available free-of-charge. Although commonly used in K–8 classrooms, Lego robots are expensive.

Preservice teachers applied their coding skills to education by creating programs appropriate for grade levels of their choice. One preservice teacher created a mathematics race in Scratch that children could win by solving problems correctly. Another programmed a geography trivia game for children to match landmarks, flags, and languages to their corresponding countries. A project with Lego Mindstorm robots used data logging to enhance the understanding of velocity as the robot moved. Models built with VPython included one of projectile motion and another of the Earth revolving around the sun. Thus, in addition to gaining computational thinking and programming skills, preservice teachers used those skills within content areas and with a pedagogical purpose.

Integration of Computational Thinking Into STEM Courses for Preservice Teachers

To engage preservice teachers in computational thinking throughout their training, we created modules for biology, physics, and geometry courses in our teacher education program. In addition to increasing computational thinking skills, the modules were designed to help future teachers apply computational thinking concepts to their understanding of science and math. To help ensure that module activities applied to real-world problems in interesting ways, students majoring in STEM and in education participated in their development.

Biology: Modeling disease spread. Introductory biology students seldom have the opportunity to build or modify scientific models, despite their importance to bioscience research (Donovan et al., 2015). An agent-based modeling platform such as NetLogo (https://ccl.northwestern.edu/netlogo/), in which each element of a system consists of hundreds to thousands of individual agents, is well suited for biology (Wilensky & Reisman, 2006). Because populations of people, animals, or plants are collections of individuals, agent-based modeling is intuitive to students. Traditional aggregate models, in which an entire population is represented by a single term in an equation, can be more difficult for beginning modelers to grasp.

The NetLogo module we created applied the susceptible-infected-recovered (SIR) model, commonly utilized by public health officials to predict the spread of disease, to simulate a mosquito-borne viral outbreak (Figure 3.2). A mosquito becomes a carrier of a virus after biting an infected person, and a susceptible person bitten by that mosquito is then also infected. Students varied the total number of people and mosquitoes, the percentage initially infected, the time period that they remain infectious, and their range of movement. As the infection spread, the simulation tracked the numbers of susceptible, infected, and recovered people in real time with a graph.

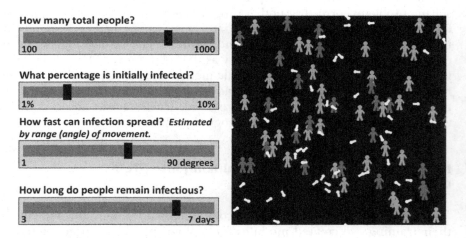

Figure 3.2.   NetLogo simulation of disease spread, with slider bars to set conditions and an interface in which people and mosquitoes move and transmit a virus.

After using the computer simulation to visualize and comprehend scenarios that allow a mosquito-borne epidemic to develop, preservice teachers expanded the model. Working in groups, they researched mosquito-borne viruses and brainstormed about ways the model could be improved. Thus, they acted as epidemiologists, who similarly start with a simple model and then add and test components. Preservice teachers' additions to the model included the possibility that female mosquitoes pass the virus to offspring and incubation periods in which people or mosquitoes had contracted the virus but were not yet infectious. Throughout this exercise, they modified computer code, developed predictions, and evaluated model results.

Finally, preservice teachers were asked to be creative and to apply the SIR model to another problem of their choice. A model of disease spread can be generalized to a wide variety of societal issues that spread from person to person, from drug abuse to fashion trends. After using decomposition and abstraction to identify the key components of their chosen problems, they developed algorithms for creating their own models with NetLogo. The preservice teachers thus practiced a computational thinking process that they can use in their future classrooms to introduce children to scientific modeling.

Physics: Visualizing vectors with interactive maps. When student attitudes towards physics are assessed in an inquiry-based, active-learning environment, the smallest gains are found in the domains of sense-making and problem-solving (Milner-Bolotin, Antimirova, Noack, & Petrov, 2011). Computational thinking offers a path to help students make sense of physics, through the problem-solving approach that is at its core. Through visualization, students can also begin to understand real-world applications of physical concepts, rather than thinking of physics as just a series of formulas and facts. Computational thinking and visualization are thus vigorously advocated by the physics education research community (Behringer & Engelhardt, 2017; Chabay & Sherwood, 2010).

The module we created employed computer simulations to visualize vectors, which are challenging for beginning physics students to understand. A vector is a quantity with both magnitude and direction, typically represented by an arrow. Vectors are relevant to physical concepts such as displacement, velocity, acceleration, and force. Our simulation used VPython to allow students to create, manipulate, and add vectors, while viewing the results on a map (see Figure 3.3). Preservice teachers added vectors with a graphical (tip to tail) method so that they could visualize and thus intuitively understand the result of vector addition. After drawing and adding vectors on a map, they also determined the displacement between various regions across the United States. They planned a trip and used realistic estimates for travel time (from flight schedules, for example) to practice other physics applications such as how to calculate average velocity.

The preservice teachers also went beyond using the simulation and learned to edit the computer code. They read and identified parts of the code that dealt with the visual aspects of the simulation, such as the color of the vectors and the explanatory text, and then edited these parts of the code. Because VPython displays the graphical output alongside the computer code, preservice teachers could immediately see if the changes they made to the visual aspects of the simulation had the intended result.

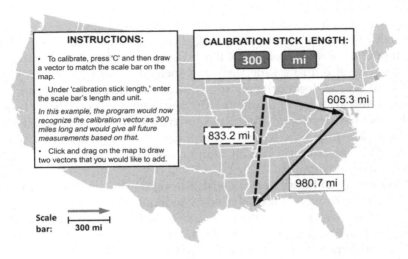

Figure 3.3. VPython simulation for adding vectors on a map.

Finally, preservice teachers uploaded maps of their own neighborhoods, places of interest, or locations involving modern-day events. Thus, they visualized vector displacements as applied to environments and neighborhoods with which they were intimately familiar, in which they had an interest, or with which they were connected through events and classes outside of physics. With the new map uploaded, the preservice teachers were asked to apply their knowledge of vectors and vector addition to discuss a new trip they created. They thus learned how data can be represented through a computer simulation and generalized the model to apply to a real-life question of interest. As a result, they will be better prepared as teachers to help children use computational thinking to solve problems.

Geometry: Visualizing and designing shapes. Students often struggle with visualizing shapes and how they are represented in the world around them. Identifying relationships between shapes and justifying their geometric properties is particularly challenging (Zambak & Tyminsk, 2017). To help preservice teachers visualize geometric shapes while also gaining computational thinking skills, we developed a module for drawing two-dimensional and three-dimensional shapes with Scratch. With its drag-and-drop programming platform, Scratch is designed to allow beginners to build programming and computational thinking skills and is an ideal tool for preservice teachers to bring to their future math classrooms (Gadanidis, Cendros, Floyd, & Namukasa, 2017).

First, preservice teachers were given a Scratch Starter Guide we developed which guided them on how to create geometric shapes with Scratch.

They began by using an algorithm to draw a 30-60-90 triangle, which they modified to build triangles of their choice, such as equilateral or isosceles triangles. They then built their own algorithms to create quadrilaterals such as squares and parallelograms, thereby generalizing and transferring a problem-solving process to a new task. For example, the code shown in Figure 3.4 executes a drawing by moving a pen a defined number of steps and by using a variable to specify the degrees for the pen to turn. They also learned coding techniques for achieving efficient solutions, such as the use of a loop to repeat a series of steps.

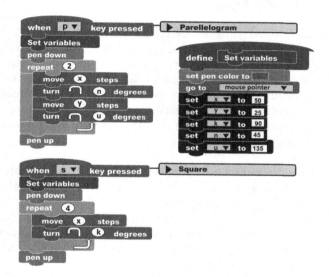

Figure 3.4   Sample code solution to create a square and a parallelogram in Scratch.

The writing of computer code to draw shapes also improves understanding of geometry concepts. For example, preservice teachers quickly recognized that it was not the interior angles of a shape that allowed it to be drawn correctly, but that the exterior angles determined the correct turn. Reusing code from previous designs and modifying it to draw more complex shapes also helped them understand the relationships between shapes. Two triangles can create a quadrilateral and three triangles a pentagon. Such visualization enabled them to imagine how geometry relates to daily life, such as in artistic designs or in planning how everything will fit into a room.

By integrating computational thinking, our module built upon the typical approach for teaching geometry. While traditional learning with paper and pencil allows for static representation of shapes, computer technology provides a dynamic environment in which to explore geometry. Preservice teachers can manipulate shapes with packages such as dynamic geometry software (Hollebrands & Lee, 2016), but our module took learning a step further by allowing flexibility for them to combine shapes, build their own designs, and explore their creations. The Scratch coding assignments not only provided a foundation in geometric principles of shapes, but also increased our preservice teachers' ability to think algorithmically, transfer problem-solving processes to new tasks, and create efficient solutions. Preservice teachers can apply these computational thinking skills as practicing educators.

Science Teaching Methods

In this capstone course, preservice teachers learned to develop lesson plans and materials for teaching science in elementary and middle school classrooms. They utilized the inquiry-based 5E instructional model to engage, explore, explain, elaborate, and evaluate. The preservice teachers also connected their lesson plans to NGSS's (n.d.) disciplinary core ideas (physical science, life science, earth & space science, and engineering), cross cutting concepts (e.g., energy, scale, systems, and patterns) and science and engineering practices (e.g., using mathematics and computational thinking).

To connect computational thinking to their future teaching, preservice teachers began by engaging with a computer simulation we developed to enhance understanding of Newton's Second Law of Motion (Adler & Kim, 2018). Although among the most important concepts in physics, Newton's laws of motion can be difficult to understand for beginning students due to misconceptions about force and motion (Bayraktar, 2009). Newton's second law describes the relationship between force, mass, and acceleration ($F = $ ma). For example, if the same net force is applied on a car and a more massive truck, the car will accelerate more than the truck because the car has less mass. Using our computer simulation readily demonstrated this relationship, because the mass and force could can be changed and the resulting acceleration immediately observed.

In addition to helping preservice teachers understand the scientific concept of motion, the computer simulation also built computational thinking skills. Preservice teachers manipulated parameters by choosing values for force and mass for two cars that competed in a race across the screen (see Figure 3.5). They made predictions as to the effect of their choices upon acceleration and evaluated if the resulting movement of the cars corresponded with their understanding. They also evaluated a graph of distance vs. time drawn in real time in the simulation as the cars moved across the screen. Finally, the preservice teachers took a quiz embedded within the simulation that gave immediate feedback on scientific knowledge gained.

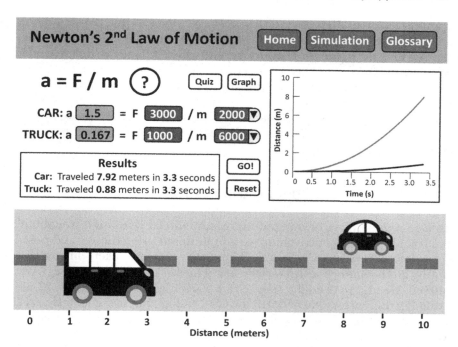

Figure 3.5. Simulation of Newton's second law with mass and force manipulated.

After playing with the simulation, preservice teachers had to stop thinking as students and start acting like teachers. They created a science lesson about a topic of their choice that included the building of concept maps (https://bubbl.us/). Concepts maps, commonly used by elementary and middle school educators, are graphical tools for identifying the important parts of a concept and illustrating the relationships between those parts. Organizing knowledge into this hierarchical framework enhances learning capability for children (Novak & Canãs, 2007). Concept mapping involves decomposing a problem into its parts, organizing information to represent relationships, and thinking algorithmically to place components in order (Mouza et al., 2017). For Newton's second law, a concept map would display the relationship between the parameters of force, mass, and acceleration with lines and arrows and include brief descriptions that abstract examples from the more complex concept (e.g., heavier car → accelerates less; lighter car → accelerates more).

Preservice teachers also developed Kahoot quizzes as part of their lesson plans (https://kahoot.it/). Kahoot is a game-based classroom response system that is played in real time with a smartphone, tablet, or computer and is a fun and interactive way for children to learn important concepts. To develop the interactive quizzes, preservice teachers represented data appropriately for different grade levels. For Newton's second law of motion, a Grade 5 student might be asked, "Which car went faster after leaving the starting line, the lighter or heavier one, and why?" A Grade 8 student could calculate the force (F) required for a car of a particular mass (m) to achieve a desired acceleration (a). The ultimate goal was for preservice teachers to be able not only to use technology to build conceptual understanding in children, but also to apply computational thinking when developing lesson plans (Mouza et al., 2017). They will then be better positioned to teach children to think computationally.

SURVEY AND RESULTS

To assess the impact of integrating computational thinking throughout the teacher training curriculum, we administered pre-semester and post-semester surveys in each course. These surveys were taken during the Spring 2018 semester, which was the first semester that all five courses were offered. We had an overall response rate of 79%, which included 14, 22, 9, 10, and 9 responses (total of 64) in the CS0, biology, physics, geometry, and science teaching methods courses, respectively. Some preservice teachers were enrolled in more than one course.

The surveys gauged preservice teachers' interest and confidence in computational thinking and in incorporating computing concepts and methods into their future teaching. The same ten questions were given at both the beginning and end of the semester (see Table 3.1). Possible answers were on a 5-level Likert scale, with strongly disagree (1), disagree (2), neutral (3), agree (4), and strongly agree (5). To compare respondents' initial and final responses to the survey questions, we conducted paired t tests. Because a total of 64 students provided responses to each of the ten questions at both the beginning and end of the semester, the number of replicates was 64 ($n = 64$) for each paired t test).

As shown in Table 3.1, for 4 of the 10 survey statements, post-semester scores were significantly higher than pre-semester scores. After completing the courses, preservice teachers rated themselves higher on statements about their knowledge and understanding of computing, technology, and computer science. They felt that they increased their abilities to think computationally, such as to use logic and algorithms and to write computer code. For example, at the beginning of the semester 66% disagreed or strongly disagreed with the statement, "I have the ability to write code for a computer program." At the end of the semester, that percentage had decreased to 28%. In response to an open-ended post-semester question, one preservice teacher wrote, "*I am comfortable to the point of being able to change a code without panicking or feeling like I am going to break it or damage it*

Table 3.1.
Survey of Perceptions of Computational Thinking

Survey Query		% Disagree (1) or Strongly Disagree (2)	% Agree (4) or Strongly Agree (5)	Mean (St. Dev.)	Paired t test
Survey statements with statistically significant increases (p < 0.05) from pre-semester to post-semester scores:					
I am knowledgeable about computing and technology.	Pre-	28%	47%	3.23 (1.07)	t = -3.00
	Post-	9%	67%	3.67 (0.84)	p = 0.004
I understand how computer scientists approach problems.	Pre-	34%	30%	2.94 (1.21)	t = -3.95
	Post-	20%	58%	3.50 (1.05)	p < 0.001
I am able to use logic or algorithms.	Pre-	20%	47%	3.39 (1.05)	t = -2.52
	Post-	13%	69%	3.72 (1.03)	p = 0.014
I have the ability to write code for a computer program.	Pre-	66%	25%	2.45 (1.34)	t = -5.23
	Post-	28%	48%	3.28 (1.15)	p < 0.001
Survey statements with no significant differences between pre-semester and post-semester scores:					
I am interested in computing and technology.	Pre-	13%	61%	3.67 (1.02)	t = -1.29
	Post-	11%	64%	3.84 (1.04)	p = 0.200
I am confident that I can learn computer science concepts.	Pre-	6%	75%	3.89 (0.86)	t = -0.12
	Post-	8%	75%	3.91 (0.90)	p = 0.902
I feel that I can relate computing or technology concepts to a variety of topics.	Pre-	19%	56%	3.52 (1.18)	t = -1.49
	Post-	8%	69%	3.78 (0.88)	p = 0.142
I am interested in using computing methods in my future classroom.	Pre-	3%	80%	4.09 (0.90)	t = 0.66
	Post-	5%	83%	4.00 (0.85)	p = 0.512
I understand computing or technology concepts well enough to incorporate them in my future classroom.	Pre-	17%	59%	3.58 (1.14)	t = -0.39
	Post-	11%	59%	3.64 (0.93)	p = 0.701
I am confident in my ability to develop the capacity to incorporate computing concepts in my teaching.	Pre-	6%	69%	3.83 (0.94)	t = 0.43
	Post-	8%	70%	3.77 (0.90)	p = 0.666

in any way. I would love to make further changes to a code if given the opportunity and time." Another stated, "*I was impressed by myself. I never expected myself to be able to generate codes and get a program to work. This made me feel proud of myself and that is what I want kids and other students to feel too: to feel accomplished.*"

Also as shown in Table 3.1, the remaining six statements showed no significant differences in pre-semester and post-semester scores. These statements assessed preservice teachers' interest and confidence in computer science, particularly in relation to incorporating computing methods and concepts into their future classrooms. Although the survey did not report significant gains for these statements at the end of the semester, both the pre-semester and post-semester scores were high. Preservice teachers largely agreed with the statements, both at the beginning and at the end of the semester. For example, at the beginning of the semester 69% agreed or strongly agreed with the statement, "I am confident in my ability to develop the capacity to incorporate computing concepts in my teaching." At the end of the semester such confidence similarly high, at 70%. One preservice teacher wrote, "*I will be able to show my students to construct figures not only by hand using a compass, but also how to construct geometric shapes using Scratch. This will teach them a little about coding and how to incorporate computer science with geometric thinking.*"

DISCUSSION AND CONCLUSION

Our modifications to the teacher education program embedded computational thinking throughout the curriculum in five courses taken by elementary and middle school preservice teachers. This comprehensive approach was more expansive than previously reported efforts to introduce computational thinking into a single course. In an educational psychology course, Yadav, Mayfield, Zhou, Hambrusch, and Korb (2014) used lecture and discussion in a 1-week unit to introduce K–12 preservice teachers to the principles of computational thinking and how to apply them to a wide variety of subjects in everyday instruction. In a science teaching methods course, preservice elementary school teachers built and programmed robots during two 3-hour labs (Jaipal-Jamani & Angeli, 2017). Participants gained both programming skills and understanding of basic science concepts, such how gears work. In an educational technology course, Mouza et al. (2017) engaged future K–8 teachers in computational thinking activities throughout a semester. Their preservice teachers gained coding experience, learned how pedagogical techniques such as concept mapping can build computational thinking skills, and integrated computational thinking into lessons they created.

By introducing computational thinking throughout our education curriculum, within multiple contexts and disciplines, we applied the principles of TPACK (Koehler et al., 2014). In our program, preservice teachers obtained a foundation in computational thinking in an introductory computer science course, built upon those skills in modules integrated into biology, physics, and geometry courses, and learned to apply computational thinking to their future teaching in a capstone science teaching methods course. Thus, they used technological knowledge within the context of content knowledge and pedagogical knowledge. The technological knowledge they gained was not confined to the use of technological tools, but focused on computational thinking (CT) skills that preservice teachers can apply to their future classrooms. This is an important distinction made by Mouza et al. (2017), who coined the term TPACK-CT. The objective is for preservice teachers to be able to foster computational thinking skills and practices among children in their future classrooms.

A strength of our program modification was the close collaboration between education, computer science, math, and science faculty in modifying our courses to include computational thinking. Yadav, Gretter, Good, and McLean (2017) and Yadav, Stephenson, and Hong (2017) argued that computer science and education faculty should work together to integrate computational thinking into educational technology courses and teaching methods courses. We extended that argument to include faculty who taught required subject area courses taken by preservice teachers. Because most K–8 teachers are generalists, preservice teachers should be provided multiple opportunities and time to apply computational thinking across a variety of contexts.

We took several intentional steps to overcome the challenges of working as an interdisciplinary team. The "silo mentality," common at universities, can inhibit interactions between departments or colleges. In regular meetings and in a 2-day workshop, we defined common goals such as desired learning outcomes for computational thinking and provided input and feedback to one another as we developed course activities. We also assembled interdisciplinary teams of students who worked with faculty throughout a summer to develop the biology, physics, and geometry modules. The teams included students majoring in science, math, computer science, and education. Each team member brought a unique perspective and skill set from his or her respective discipline while the students and the faculty helped one another to learn the coding skills, science and math content, and implement the educational pedagogy necessary to create each module. Throughout this project, we benefitted from administrative support and from external funding, which allowed us to provide stipends to faculty and students.

The results of our multidisciplinary approach to engage preservice elementary and middle school teachers in computational thinking are promising, but mixed. In the post-semester survey, preservice teachers rated themselves significantly higher in their knowledge of computer science and in their abilities to think computationally, such as to use algorithms and logic and to write computer code. However, there was no change in their interest and confidence in incorporating computing concepts and methods into their future classrooms. This may be partly because most had not at that stage taken science teaching methods, in which they learn to integrate computational thinking into lesson plans. We plan to further emphasize the application of computational thinking to K–8 education throughout each course so that preservice teachers do not wait until the capstone course to start thinking about how to apply computational thinking to their future teaching.

Mouza et al. (2017) similarly reported a disconnect between understanding of computational thinking and application to teaching. After building and programming robots in an educational technology course, their pre-course and post-course surveys showed significant gains in K–8 preservice teachers' understanding and knowledge of computational thinking. However, their preservice teachers' comfort with and interest in computational thinking, potential classroom use, and the perceived importance of computational thinking toward future career goals did not significantly change. Moreover, while some of their preservice teachers were able to integrate computational thinking concepts and tools into a lesson plan, others struggled to relate computational thinking to disciplinary content and pedagogy in a meaningful way. Also, despite gaining practice with programming tools such as Scratch, none of their preservice teachers chose to include programming in their lesson plans, instead relying on predesigned software applications.

Misconceptions at the beginning of the semester may also explain why exposure to computational thinking in our courses did not increase interest in incorporating it into future teaching. Yadav et al. (2014) and Yadav, Gretter, et al. (2017) found that preservice teachers with no previous exposure to computational thinking had a wide range of ideas about what the term meant and that many exhibited an oversimplified view of computational thinking concepts and practices. They commonly conflated computational thinking with just the use of computers or technology, or with other types of thinking such as mathematical thinking or logical thinking. Consequently, after participating in a computational thinking unit, preservice teachers may realize how much they do not know. In our study, it is possible that a lack of understanding of computational thinking at the beginning of the semester led to overconfidence during the pre-semester survey when participants were asked to assess their understanding of and confidence in using computing concepts and methods. Although over-confidence may help explain the initially high scores in these areas, it is important to note that final scores remained high. In other words, pre-service teachers began the semester with high interest and confidence in computing and in incorporating computing concepts and methods into their future teaching, and maintained that interest and confidence after learning about computational thinking in our courses.

As our program continues, we will follow preservice teachers as they complete all five courses that integrate computational thinking and we are expanding our assessment efforts. To further assess knowledge of computational thinking, we are adding questions to the pre-semester and post-semester surveys about specific concepts. For example, we will ask about participants' comfort with breaking down a problem into smaller parts so that a computer can solve it and about their understanding of how computers can be used to model phenomena. We will also assess preser-vice teachers' computational thinking skills with a rubric we developed to evaluate performance on class activities and assignments. Congruent with Bloom's taxonomy (Anderson et al., 2001), the rubric assesses abilities to understand and apply computational thinking to solve a problem, analyze and evaluate the results obtained, and to use that experience to create solutions to new problems.

In-service teachers, particularly those with little computer science experi-ence, also benefit from training in computational thinking (Israel, Pearson, Tapia, Wherfel, & Reese, 2015). Going forward, we hope to broaden our program to include in-service teachers who would like to integrate compu-tational thinking into their classrooms. We have already started such efforts by offering workshops attended by current teachers pursuing continuing education (Mayle & Rabe, 2018; Slate, 2018a, 2018b). We also plan to follow our graduates as they matriculate into teaching, through a long-standing master teacher designation that keeps interested alumni connected to our teacher education program. This will give us an opportunity to learn of their experiences and challenges with incorporating computational think-ing into the K–8 schools at which they are teaching.

Our revised teacher training program, which integrates computational thinking throughout the curriculum, has begun successfully. Preservice teachers began the semester with high interest and confidence in computing methods and technology, and in incorporating computing concepts into their future classrooms. At the end of the semester, they had similarly high interest and confidence, despite possibly realizing that their initial understanding of computational thinking may have been incomplete. They also reported statistically significant gains in their perceived knowledge of computing and computer science and in their perceived computational thinking skills. These results came from close collaboration between education, computer science, math, and science faculty as we incorporated computational thinking into courses taken by future elementary and middle school teachers. Computational thinking was integrated with discipline-specific content and with the teaching of pedagogy, through a three-tiered approach that included a foundational computer science course, three STEM content courses, and a capstone science teaching methods course. By updating our preservice teaching curriculum in these five courses, we are providing future teachers with a comprehensive foundation in computational thinking that they can use as educators.

ACKNOWLEDGMENTS

Students Uzma Ain, Estefania Figueroa, Daniel Fitch, Emmet Hilly, Jonathan Jurczak, Michael Konecki, Lauren Rabe, Itzel Ruiz, Jeremiah Santos, Samah Slim, Shane Taylor, and Stuart Thiel were instrumental in the creation and testing of the computational thinking modules. Brittany Pines and Dr. Durene Wheeler provided valuable help in advising and guiding the students with a particular emphasis on how to work in diverse teams and ensuring many of the course modules we created had cultural and real-world relevance. We thank an anonymous reviewer for constructive feedback and suggestions.

This material is based upon work was supported by the National Science Foundation under Grant No. DRL-1640041. Any opinions, findings, and conclusions or recommendations expressed in this material are those of the authors and do not necessarily reflect the view of the National Science Foundation.

REFERENCES

Adler, R. F., & Kim, H. (2018). Enhancing future K–8 teachers' computational thinking skills through modeling and simulations. *Education and Information Technologies, 23*(4), 1501–1514. https://doi.org/10.1007/s10639-017-9675-1

Anderson, L. W., Krathwohl, D. R., Airasian, P. W., Cruickshank, K. A., Mayer, R. E., Pintrich, P. R., ...Wittrock, M. C. (Eds.). (2001). *A taxonomy for learning, teaching, and assessing: A revision of Bloom's taxonomy or educational objectives.* New York, NY: Longman.

Anewalt, K. (2008). Making CS0 fun: An active learning approach using toys, games and Alice. *Journal of Computing Sciences in Colleges, 23*(3), 98–105.

Barr, D., Harrison, J., & Conery, L. (2011). Computational thinking: A digital age skill for everyone. *Learning & Leading with Technology, 38*(6), 20–23. Retrieved from http://www.learningandleading-digital.com/learning_leading/20110304?pg=22#pg22

Bayraktar, S. (2009). Misconceptions of Turkish pre-service teachers about force and motion. *International Journal of Science and Mathematics Education 7*, 273–291. https://doi.org/10.1007/s10763-007-9120-9

Behringer, E., & Engelhardt, L. (2017). AAPT Recommendations for computational physics in the undergraduate physics curriculum, and the Partnership for Integrating Computation into Undergraduate Physics. *American Journal of Physics, 85*, 325–326. https://doi.org/10.1119/1.4981900

Bers, M. U., Flannery, L., Kazakoff, E. R., & Sullivan, A. (2014). Computational thinking and tinkering: Exploration of an early childhood robotics curriculum. *Computers & Education, 72*, 145–157. https://doi.org/10.1016/j.compedu.2013.10.020

Calder, N. (2010). Using Scratch: An integrated problem-solving approach to mathematical thinking. *Australian Primary Mathematics Classroom, 15*(4), 9–14.

Chabay, R., & Sherwood, B. (2010) *Matter and interactions* (3rd ed.). Somerset, NJ: John Wiley.

Cliburn, D. C. (2006). A CS0 course for the liberal arts. *ACM SIGCSE Bulletin, 38*(1), 77–81. https://doi.org/10.1145/1124706.1121368

Cortina, T. (2007). An introduction to computer science for non-majors using principles of computation. *ACM SIGCSE Bulletin. 39*(1), 218–222. https://doi.org/10.1145/1227310.1227387

Donovan, S., Eaton, C. D., Gower, S. T., Jenkins, K. P., LaMar, M. D., Poli, D., ... Wojdak, J. (2015). QUBES: A community focused on supporting teaching and learning in quantitative biology. *Letters in Biomathematics, 2*(1), 46–55. https://doi.org/10.1080/23737867.2015.1049969

Gadanidis, G., Cendros, R., Floyd, L., & Namukasa, I. (2017). Computational thinking in mathematics teacher education. *Contemporary Issues in Technology and Teacher Education, 17*(4), 458–477. Retrieved from https://www.citejournal.org/publication/volume-17/issue-4-17/

Grover, S., Pea, R., & Cooper, S. (2014). Remedying misperceptions of computer science among middle school students. In J. D. Dougherty, K. Nagel, A. Decker, & K Eiselt (Eds.), *Proceedings of the 45th ACM Technical Symposium on Computer Science Education* (pp. 343–348), Atlanta, GA: ACM. https://doi.org/10.1145/2538862.2538934

Grubbs, M. (2013). Robotics intrigue middle school students and build STEM skills. *Technology and Engineering Teacher, 76*(6), 12–16.

Hollebrands, K., & Lee, H. (2016). Characterizing questions and their focus when pre-service teachers implement dynamic geometry tasks. *The Journal of Mathematical Behavior, 43*, 148–164. https://doi.org/10.1016/j.jmathb.2016.07.004

International Society for Technology in Education (ISTE) and The Computer Science Teachers Association (CSTA). (2011). *Operational definition of computational thinking for K–12 education.* Retrieved from http://www.iste.org/docs/ct-documents/computational-thinking-operational-definition-flyer.pdf?sfvrsn=2

Israel, M., Pearson, J. N., Tapia, T., Wherfel, Q. M., & Reese, G. (2015). Supporting all learners in school-wide computational thinking: A cross-case qualitative analysis. *Computers & Education, 82,* 263–279. http://dx.doi.org/10.1016/j.compedu.2014.11.022

Jaipal-Jamani, K., & Angeli, C. (2017). Effect of robotics on elementary preservice teachers' self-efficacy, science learning, and computational thinking. *Journal of Science Education and Technology, 26*(2), 175–192. http://doi.org/10.1007/s10956-016-9663-z

Koehler, M. J., Mishra, P., Kereluik, K., Shin, T. S., & Graham, C. R. (2014). The technological pedagogical content knowledge framework. In J. M. Spector, M. Merrill, J. Elen, & M. Bishop (Eds.), *Handbook of research on educational communications and technology* (pp. 101–111). New York, NY: Springer. https://doi.org/10.1007/978-1-4614-3185-5_9

Lee, I., Martin, F., Denner, J., Coulter, B., Allan, W., Erikson, J., Malyn-Smith, J., & Werner, L. (2011). Computational thinking for youth in practice. *ACM Inroads, 2*(1), 32–37. http://doi.org/10.1145/1929887.1929902

Leonard, A. E., & Daily, S. B. (2013). The dancing Alice project: Computational & embodied arts research in middle school education. *Voke, 1.* Retrieved from http://www.vokeart.org/?p=331&spoke=1

Leonard, J., Buss, A., Gamboa, R., Mitchell, M., Fashola, O. S., Hubert, T., & Almughyirah, S. (2016). Using robotics and game design to enhance children's self-efficacy, STEM attitudes, and computational thinking skills. *Journal of Science Education and Technology, 25*(6), 860–876. https://doi.org/10.1007/s10956-016-9628-2

Maloney, J. H., Peppler, K., Kafai, Y., Resnick, M., & Rusk, N. (2008). Programming by choice: Urban youth learning programming with scratch. *ACM SIGCSE Bulletin, 40*(1), 367–371. https://doi.org/10.1145/1352322.1352260

Mayle, S., & Rabe, L. (2018, April). *Using physics modules to incorporate computational thinking into teacher education in STEM.* Paper presented at the 20th Annual Chicago Symposium Series: Excellence in Teaching Mathematics and Science, Chicago, IL.

Milner-Bolotin, M., Antimirova, T., Noack, A., & Petrov, A. (2011). Attitudes about science and conceptual physics learning in university introductory physics courses. *Physical Review Special Topics-Physics Education Research 7*(2), 020107(1–7). https://doi.org/10.1103/physrevstper.7.020107

Mishra, P., & Koehler, M.J. (2006). Technological pedagogical content knowledge: A framework for teacher knowledge. *Teachers College Record, 108*(6), 1017–1054. Retrieved from https://www.tcrecord.org/content.asp?contentid=12516

Mouza, C., Yang, H., Pan, Y.-C., Yilmaz Ozden, S., & Pollock, L. (2017). Resetting educational technology coursework for pre-service teachers: A computational thinking approach to the development of technological pedagogical content knowledge (TPACK). *Australasian Journal of Educational Technology, 33*(3), 61–76. https://doi.org/10.14742/ajet.3521

National Research Council. (2012). *A framework for K–12 science education: Practices, crosscutting concepts, and core ideas.* Washington, DC: National Academies Press. https://doi.org/10.17226/13165

Next Generation Science Standards Science and Engineering Practices. (n.d.) Retrieved from https://ngss.nsta.org/PracticesFull.aspx

Novak, J. D., & Canãs, A. J. (2007). Theoretical origins of concept maps, how to construct them, and uses in education *Reflecting Education, 3*(1), 29–42. Retrieved from http://www.reflectingeducation.net/index.php/reflecting/article/view/41

Ornek, F. (2008). Models in science education: Applications of models in learning and teaching science. *International Journal of Environmental & Science Education, 3*(2), 35–45. Retrieved from http://www.ijese.net/arsiv/8

Powers, K., Ecott, S., & Hirshfield, L. M. (2007). Through the looking glass: Teaching CS0 with Alice. *ACM SIGCSE Bulletin, 39*(1), 213-217. https://doi.org/10.1145/1227504.1227386

Resnick, M., Maloney, J., Monroy-Hernández, A., Rusk, N., Eastmond, E., Brennan, K., . . . Kafai, Y. (2009). Scratch: programming for all. *Communications of the ACM, 52*(11), 60–67.

Rizvi, M., Humphries, T., Major, D., Jones, M., & Lauzun, H. (2011). A CS0 course using Scratch. *Journal of Computing Sciences in Colleges, 26*(3), 19–27.

Sengupta, P., Kinnebrew, J. S., Basu, S., Biswas, G., & Clark, D. (2013). Integrating computational thinking with K-12 science education using agent-based computation: A theoretical framework. *Education and Information Technologies, 18*(2), 351–380. https://doi.org/10.1007/s10639-012-9240-x

Settle, A., Franke, B. & Hansen, R., Spaltro, F., Jurisson, C., Rennert-May, C., & Wildeman, B. (2012). Infusing computational thinking into the middle- and high-school curriculum. In T. Lapidot, J. Gal-Ezer. M. E. Caspersen, & O. Hazzan (Eds.), *Proceedings of the 17th ACM Annual Conference on Innovation and Technology in Computer Science Education* (ITiCSE '12) (pp. 22–27). New York, NY: ACM. https://doi.org/10.1145/2325296.2325306

Slate, J. E. (2018a, April). *Incorporating modeling and computer coding into science and math courses.* Paper presented at the 20th Annual Chicago Symposium Series: Excellence in Teaching Mathematics and Science, Chicago, IL.

Slate, J. E. (2018b, June). *Scientific modeling: Not just for scientists!* Paper presented at the Inquiry and Design Institute for Connected Learning, Center for College Access and Success, Northeastern Illinois University, Chicago, IL.

Uludag, S., Karakus, M., & Turner, S. W. (2011). Implementing IT0/CS0 with Scratch, app inventor for Android, and Lego Mindstorms. In B. Goda, E. Sobiesk, & R. Connolly (Eds.), *Proceedings of the 2011 Conference on Information Technology Education* (SIGITE '11) (pp. 183–190). New York, NY:ACM. https://doi.org/10.1145/2047594.2047645

Voogt, J., Fisser, P., Good, J., Mishra, P., & Yadav, A. (2015). Computational thinking in compulsory education: Towards an agenda for research and practice. *Education and Information Technologies, 20*(4), 715–728. https://doi.org/10.1007/s10639-015-9412-6

Wang, J. (2017). Is the U.S. education system ready for CS for all? *Communications of the ACM, 60*(8), 26–28. https://doi.org/10.1145/3108928

Wilensky, U., & Reisman, K. (2006). Thinking like a wolf, a sheep, or a firefly: Learning biology through constructing and testing computational theories-an embodied modeling approach. *Cognition and Instruction, 24*(2), 171–209. https://doi.org/10.1207/s1532690xci2402_1

Wing, J. M. (2006). Computational thinking. *Communications of the ACM, 49*(3), 33–35. https://doi.org/10.1145/1118178.1118215

Wing, J. M. (2008). Computational thinking and thinking about computing. *Philosophical Transactions of the Royal Society A: Mathematical, Physical and Engineering Sciences, 366*(1881), 3717–3725. https://doi.org/10.1098/rsta.2008.0118

Wolz, U., Stone, M., Pearson, K., Pulimood, S. M., & Switzer, M. (2011). Computational thinking and expository writing in the middle school. *ACM Transactions on Computing Education, 11*(2). http://dx.doi.org/10.1145/1993069.1993073

Yadav A., Gretter S., Good J., & McLean T. (2017) Computational thinking in teacher education. In P. Rich & C. Hodges (Eds.), *Emerging research, practice, and policy on computational thinking* (pp. 205–220). Springer, Cham. https://doi.org/10.1007/978-3-319-52691-1_13

Yadav, A., Mayfield, C., Zhou, N., Hambrusch, S., & Korb, J. T. (2014). Computational thinking in elementary and secondary teacher education. *ACM Transactions on Computing Education, 14*(1), 1–16. http://dx.doi.org/10.1145/2576872

Yadav, A., Stephenson, C., & Hong, H. (2017). Computational thinking for teacher education. *Communications of the ACM, 60*(4), 55–62. https://doi.org/10.1145/2994591

Zambak, V. S., & Tyminski, A. M. (2017). A case study on specialised content knowledge development with dynamic geometry software: The analysis of influential factors and technology beliefs of three pre-service middle grades mathematics teachers. *Mathematics Teacher Education and Development, 19*(1), 82–106. Retrieved from https://mted.merga.net.au/index.php/mted/article/view/311

CHAPTER 4

THE AVAILABILITY OF PEDAGOGICAL RESPONSES AND THE INTEGRATION OF COMPUTATIONAL THINKING

Whitney Wall Bortz, Aakash Gautam, Deborah Tatar
Virginia Tech

Kemper Lipscomb
University of Texas

Rapid technological advances have made computational skills more important toc fully participate in society and our global economy, and as a result, computational thinking has earned attention in K–12 education as a means to preparing the future workforce. With an already congested curriculum, one approach is to integrate computational thinking into core classes that all students already take. This chapter reports on the integration of computational thinking in middle school chemistry classes. We focus on one teacher's pedagogical strategies employed in each of the two integrated domains—in this case, computational thinking and science. We describe variations in the teacher's approach across the two domains, noting that certain approaches to pedagogy and classroom discourse align with computational thinking. Such strategies were observed less frequently when the instruction focused

Integrating Digital Technology in Education:
School–University–Community Collaboration, pp. 81–109
Copyright © 2019 by Information Age Publishing
All rights of reproduction in any form reserved.

on the less familiar domain of computational thinking. Second, it cannot be sufficiently emphasized that programming and computational thinking are entirely new fields for most teachers and teachers need considerable opportunity to learn this new domain to optimize its integration.

This chapter describes one teacher's pedagogical variation across two integrated domains: science and computational thinking (CT). We show how when teaching science, the teacher engages students and uses discourse that promotes the development of CT. However, instruction of programming-related concepts, which are new to the teacher, is more teacher-centered. The example of one teacher's varied approaches to teaching the two domains highlights the role of pedagogical content knowledge and its effect on the employment of best pedagogical practices for an integrated CT environment. The data presented here are part of a larger study consisting of three curricular units that integrate traditional approaches to science instruction—such as experimentation, lecture and discussion—with activities embedded in three different computer simulations. Investigating pedagogical approaches which integrate technology is timely given the trend toward technology adoption in schools and in particular the movement towards CT.

The 2017 NMC Horizon Report (Freeman, Adams Becker, & Cummins, 2017) identified 10 "Big Picture Themes" of educational change present in the movement towards technology and CT integration. Central to this chapter is the theme, "there is no replacement for good teaching" (p. 4). Teachers now have access to a plethora of technologies that hold potential for enhancing student learning, but their use of those resources in the classroom matters. Teachers must rethink the content and structure of their instruction and assessment in light of technology integration (Clemons, 2006; De Freitas & Neumann, 2009; Yadav, Mayfield, Zhou, Hambrusch, & Korb, 2014). The teacher is an important factor through whom the learning is mediated. Their pedagogical decisions interact with the technology's potential as a learning medium, particularly in the case of integrating CT into core curriculum. This argument must be made in part because, despite the importance of technology and CT, there is little research on how teachers actually incorporate new technologies into their teaching (Puttick, Drayton, & Karp, 2015).

In this chapter, drawing on a sociocultural lens, we compare one teacher's instruction when the focal domain is science with his instruction that centers on computational thinking and computer science (CT/CS). His science instruction is student-centered, involving students in classroom discourse. He uses evidenced-based inquiry practices such as skillful questioning techniques, scaffolding student learning, and guiding students as they do the higher order cognitive work. In the context of integrated CT, the teacher's discourse actually aligns with ideas of CT, thus making the CT more accessible, we argue, to the students. Analyses of his teaching practices led us to two main arguments presented here. First, in the integrated CT classroom, successful integration rests on more than the technology itself or even the accompanying curricular design. Even when focusing on the primary domain (science), the teacher's specific pedagogical moves and ways of speaking can complement students' development of CT. Second, when the teacher's focus shifts to the CT, a lack of CT domain expertise and pedagogical content knowledge can create instructional challenges, making the potential of this integrated approach difficult to gauge.

THEORETICAL UNDERPINNINGS

Computational Thinking and Its Integration

Cuny, Snyder, and Wing (2010) define CT as the thought process that involves "formulating problems and their solutions so that the solutions are represented in a form that *can be effectively carried out by an information-processing agent*" [emphasis added]—a skill that is widely accepted as essential for the holistic education of the future workforce (National Research Council, 2010). While it is argued that students should receive opportunities to develop CT early on in their K–12 experience (Grover & Pea, 2018; Israel, Pearson, Tapia, Wherfel, & Reese, 2015; Lye & Koh, 2014), a congested curriculum prevents universal provision of explicit instruction in programming and CT skills (Qualls & Sherrell, 2010; Sengupta, Kinnebrew, Basu, Biswas, & Clark, 2013; Settle et al., 2012; Wilensky & Reisman, 2006). On the other hand, if CT/CS is offered only as an optional elective underserved students often either self-select out or are pushed out by structural and institutional barriers. An alternative approach is to integrate computational activities into core classes that all students take, with the aim of deepening learning in the core discipline while also introducing students to computational tools and skills (diSessa, 2000; Kaput & Schorr, 2008; Papert, 1980; Wilensky & Stroup, 1999). Therefore, integrating CT into STEM is an avenue for building computational fluency in *ALL* students (Grover & Pea, 2018; Qualls & Sherrell, 2010).

The middle school (MS) science classroom lends an ideal space for this, as it is a discipline required for all students, and natural connections exist between CT and science pedagogy (Dickes & Sengupta, 2013; Goldstone & Wilensky, 2008; Jacobson, 2006; Reed et al., 2005). Still, the existence of varied perspectives on both the nature of CT (Grover & Pea, 2018; National Research Council, 2010) as well as on how to integrate it into science learning (National Research Council, 2011) adds additional constraints, challenges, and issues with cognitive load that must inform our development of best pedagogical approaches.

Learning and Pedagogy

Wenger's (1998) model of learning in communities of practice as well as other sociocultural approaches served as key guides to the analyses of pedagogical practices in this chapter. Communities as well as individuals change through the incorporation of new practices. In the context of this research, the research team, the teachers, and the participating students can all be considered participants in communities that are in flux, each with varying starting points and levels of sophistication. Learning is seen as a negotiation of meaning between prior understanding, the environment, and social interactions (Lave & Wenger, 1991). Thus, in our research, participants' prior experiences, tools and activities in the instructional units, and the resulting social interactions all contributed to the learning that took place. Likewise, limitations at any of these levels could impede learning.

From a sociocultural constructivist framework (Bonk & King, 1998) focus on the role of social interaction and language in learning informs the analyses of pedagogical practice in this chapter. It is through linguistic engagement and interaction that learners can experience metacognitive exercises that bring their own knowledge and abilities to the fore (Murphy, 1996); thus, discourse in the classroom becomes both a learning tool and a tool that helps make student thinking visible as well as an avenue for formative assessment of the students and the project.

Our intervention was implemented in classrooms already governed by their own unique cultures, as each classroom community has negotiated its own ways of being, acting, and communicating (Bruner, 1996; Cobb & Bowers, 1999). Moreover, each academic discipline can be considered to have its own culture (Murphy, 2008) in which students progress to operate more effectively as they encounter and adopt its associated language and tools (Bruner, 1996; Greeno, 1989; Mascolo, 2009; Murphy, 1996; van Oers, 2000; Vygotsky, 1978; Wenger, 1998). This is particularly relevant in the integrated context where the aim is to deepen student learning in two disciplines in such a way that enhances the learning of both. In spite of a shared classroom culture, each student had varying levels of sophistication in his or her ability to utilize the language and tools associated with both the science and the CT prior to the instruction of the integrated CT/science units.

THE CHEM-C PROJECT

The work reported here is part of a larger, multiyear, multischool study situated in middle school (seventh and eighth grade) classes. The intervention included three, one-week integrated replacement units in low to moderate socioeconomic status (SES) public middle schools in Texas and Virginia. Each unit combined specially designed technological environments implemented in NetLogo (Wilensky, 1999) and accompanying curricular sequences (Table 4.1). Enactment began with a partnership between university researchers and K–12 educators. Prior to implementation, teachers participated in 6 days of hands-on professional development facilitated by the research team. While the researchers led the development of the technology and curricular materials, teacher experience with and input into the materials influenced modifications. In the first year of the project, the researchers implemented the intervention in the teachers' classrooms. This pilot stage, including feedback from participating teachers, influenced refinements of the technology and curricular approach while also acting as a model for teachers who had no prior experience integrating CT in their classes. In the second year, the participating teachers autonomously led the intervention in their own classrooms while researchers observed and collected video data. Although the teachers were able to watch the researchers implement the units in the first year, the inherently connected (Lye & Koh, 2014) domains of programming and computational thinking remained new to them, and we believe their domain knowledge influenced their pedagogical execution of the curricular units.

Curriculum and Technology

Each of the three curricular units, referred to as Computational Chemistry Tasks (CCTs), was aligned to the Next Generation Science Standards (NGSS) (NGSS Lead States, 2003) and the Texas Essential Knowledge and Skills (TEKS) for science (Texas Education Agency, 2013). Each addressed chemistry concepts documented in science education literature as difficult for students to learn, such as the nature of matter (Herrmann-Abell & DeBoer, 2011; Lee, Eichinger, Anderson, Berkheimer, & Blakeslee, 1993; Stavy, 1991) and chemical equilibrium (Sendur, Toprak, & Pekmez, 2010). The computing environment for each CCT had two faces: (1) students experienced and investigated a scientific phenomenon through an animated simulation with interactors (sliders, buttons) and instruments (readouts, graphs); (2) The simulation could be explored and changed through reading and revising the code model. The approach assumed that neither students nor teachers had prior experience with CT. The general focus was

Table 4.1

CCT1 – Scope and Sequence

Guiding Question: What is happening that we do not normally see? What is present but is not shown by the model?

TEKS: 8.5 d – f

NGSS: PS1.A, PS1.B, PS3.B, MS-PS1-5, MS-PS1-4, MS-PS1-2, MS-PS1-5, MS-PS1-2

Day	Stage	Time (min)	Materials Needed	Specific Activities	Outcomes
1	1. Preassessments	50 min. or less	CAT and PACT	Collect demographic information and administer the preassessments	**Goal:** Obtain data Artifacts: questionnaire and two assessments.
2	2. Introduce an anchoring phenomenon (Schwarz et al., 2009) **Driving Question:** *What happens during this phenomenon?*	20	Beakers • Sets of two test tubes • 9-volt batteries • Epsom salt • Water • Stirring rod	• Investigate prior knowledge of composition of **matter,** what happens during a **chemical reaction,** and properties of a **water molecule.** • Students perform the experiment	**Goals:** Review and provide the visual **Artifact:** Teacher-created white board with the reaction and with input from students

	Time	Materials	Steps	Goals/Artifacts
3. Elicit ideas about models in science **Driving Question:** *What is happening that we can't see?*	20	• One white board or poster paper per student group • Markers for each group	• Teacher introduces modeling • Provide an example of a model (experiment) • Students draw the experiment. • Students share their models. • Teacher provides feedback • Discuss what was happening that could not be seen. • If time permits, students have the opportunity to revise models as a result of prior sharing/discussion.	**Goals:** Students should understand the various types and purposes of models; Students explore ideas of macro- and micro-phenomena in science; Draw connections between stoichiometry and a visual phenomenon **Artifacts:** Group white board models
4. Concluding today and looking ahead **Driving Question:** *How are computer models useful?*	10		• Teacher explains that models up to now have been static, but we will now transition to exploration of a dynamic model • Discuss advantage of a computer model (previews next lesson).	Goal: Establish the purpose of a computer model by defining the ways in which it is helpful while still being imperfect.
5. Explore an initial simulation of the phenomenon **Driving Question:** *What things do you notice happening in this model?*	15	• Laptops or Chromebooks • The "Water Forming and Splitting" Netlogo simulation	• Students explore the simulation • Teacher asks student to experiment with the components in the interface by adjusting various settings and noting what happens when running the simulation.	Goals: Engagement in investigative processes. Artifacts: Logged data from students' engagement with the simulation

(Table continues on next page

Table 4.1
(Continued)

Guiding Question: What is happening that we do not normally see? What is present but is not shown by the model?

TEKS: 8.5 d – f

NGSS: PS1.A, PS1.B, PS3.B, MS-PS1-4, MS-PS1-2, MS-PS1-5, MS-PS1-2

Day	Stage	Time (min)	Materials Needed	Specific Activities	Outcomes
3	6. Critique the initial computational model **Driving Questions:** *What is being represented in the simulation? What is scientifically accurate/inaccurate?*	15	• Science Fact Sheet (Appendix #) • Design Component Chart (DCC) (Appendix #) • "Assessing the Model" worksheet (Appendix #)	• Teacher distributes a science fact sheet as a conceptual reference. • Students identify the components of the model. Students note scientifically accurate and inaccurate components. • Students encouraged to critique • Students report out. Teacher facilitates and leads students toward a decision on the change to implement in the code.	Goals: Students learn about model components; think critically to evaluate the model scientifically Artifacts: "Assessing the Model" worksheets, Whole-class generated white boards
	7. Learning about code Intro to code **Driving Question:** *How are such models developed, and how do these models work?*	20	• Individual computers with access to the Netlogo model • Programming Lesson One (Appendix #)	• Students are guided through a series of exercises in the command center which introduce concepts such as: show/ask; change size/ change color; change xcor; ycor; change all turtles; and turtle types. • Some of these exercises include recording discoveries in the programming lesson handout	Goal: Introduction to programming concepts Artifacts: student responses on the worksheet and logged data

	Activity	Time	Materials	Description	Goals / Artifacts
4	8. Review **Driving Question:** What models have we learned about? Why are models useful?	5		• Teacher wraps up with a whole-class discussion about the big ideas of science and modeling.	
	9. Changing the Computational Code **Driving Question:** *How should we change the Netlogo code?*	45	• Individual computers with access to the Netlogo model • Programming Lesson Two (Appendix #)	• Permanent changes must happen in the code • This lesson presents new concepts, such as: variables; brackets; and code changes and guides students through how to add an object (Salt) to the simulation. • Working in groups, students make modifications to the model, test, and refine.	Goals: Students learn additional programming concepts and skills Artifacts: Responses on guided handout and logged data
5	10. Students finish changing the Netlogo code	15	• Same as above	• Students finish coding the creation of salt in the model as a group.	Continued from above

(Table continues on next page

Table 4.1
(Continued)

Guiding Question: What is happening that we do not normally see? What is present but is not shown by the model?

TEKS: 8.5 d – f

NGSS: PS1.A, PS1.B, PS3.B, MS-PS1-5, MS-PS1-4, MS-PS1-2, MS-PS1-5, MS-PS1-2

Day	Stage	Time (min)	Materials Needed	Specific Activities	Outcomes
5	11. Building and presenting an argument **Driving Question:** *How can you justify the changes that you made using both scientific and computational principles?*	25	• Individual white boards or poster paper • Markers	• Students present their changes to the class with reasoning behind the object added and the properties given to the object. • Students also share where these changes were placed in the code and why.	Goals: Students can formulate evidence-based arguments that draw upon an understanding of the role of Epsom salt in the chemical reaction as well as of objects and their properties in computer models. Artifacts: Group-generated white boards and computer logs
	12. Feedback	10	• Index cards	• Students answer 3 questions on index cards about their experiences during this CCT: 1. *What did you learn about science?* 2. *What did you learn about computers/programming?* 3. *What would you have changed about the past 5 days?*	Goal: Gather students' opinions about their own learning over the course of the CCT Artifact: Index cards

on modeling and representations; however, the mechanisms used to gain the interest of the science teachers and students was the idea of making improvements in the models to investigate scientific ideas. Programming practices were introduced to enable students to create improvements. The first idea, however, was to introduce the idea that "computer programs can be used to create the simulations," which therefore can be programmed differently.

Five experienced science teachers (four eighth grade teachers and one seventh grade teacher) acquired training through 6 days of workshops that walked them through the student experience and engaged them with ideation. Participating teachers were experienced science educators, but they had no prior experience with CT or with integrating simulations into their teaching. Therefore, we prioritized teacher learning at the start of the project. In a series of face-to-face professional development sessions led by the research team, teachers participated in assorted learning activities including direct instruction on basic programming, features of the NetLogo environment, hands-on exploration of the initial drafts of simulations, collaborative discussions and reflections about potential science learning for students, and programming tasks including modifying and building models. In the pilot stage of the project, researchers taught the CCTs in participating teachers' classes thus modeling instruction for the teachers and working alongside teachers to troubleshoot or to modify instruction in real time in response to students' needs.

Materials

In addition to the simulation environment and the code present in the simulation environment, teacher materials included a scope-and-sequence document (Table 4.1) for each CCT. This document provided focus not only on the standards addressed and the relevant activities but also on the novel pedagogical purposes of the CCT. These were present in a "driving science question," a "driving CT question," and through the identification of related skills. Teachers were expected to follow the general sequence and to collect completed student materials but were free to incorporate the tools according to their own pedagogical preferences. Student materials included both those given to the students and those created by the students. They engaged in activities supported by a structured design critique document and also a code planning document.

Instruction of each CCT, as described in the scope-and-sequence document, always started with an *anchoring phenomenon* (Schwarz et al., 2009) of the science to be modeled and understood in the unit. It moved instruction through the creation of student interpretations and group creation and critique of student generated models of the anchoring phenomenon, initial exploration of the computer simulation, student interpretation and critique of the simulation, introduction of computing practices, and student improvements to the computer simulation.

A major project decision consisted of the choice to create science simulations in NetLogo that reflected the complexity of what students were learning in science. This gained the science teachers' interests but increased challenges in the management of programming-related CT.

This chapter focuses on data gathered from the first of 3 week-long CCTs, as this exemplifies the teacher's initial attempts to integrate CT into science prior to receiving any feedback from the researchers observing or from students about their experiences. Obstacles present in the first CCT can determine the potential for adoption of an innovation (Rogers, 1995). During the first CCT the learning curve is steepest for both teachers and students.

Plan of CCT1

CCT1 was concerned with the (in theory) reversible reaction that can create hydrogen (H_2) and oxygen (O_2) from water (H_2O). The anchoring phenomenon was a physical experiment of water splitting into hydrogen and oxygen molecules in the presence of a battery and a catalyst, Epsom salt ($MgSO_4$). In small groups, students submerged a battery into a tub with a solution of Epsom salt and water, placing a test tube directly above each terminal of the battery to collect gas bubbles generated by the reaction. Students watched as bubbles formed in each test tube, one producing hydrogen gas and the other, oxygen. In groups students then created models on whiteboards—this two-dimensional medium limits the creation to *static models*—in response to the question, "What did you see happening?," thus offering an opportunity to discuss interpretations of their observations. Students were then introduced to a simulation—thus introducing the idea of a *dynamic model*— that depicted splitting of water into hydrogen and oxygen in the presence of electricity (Figure 4.1). In addition, the model depicts formation of water when hydrogen and oxygen with high energy collide. The exploration of the simulation was partly guided by a graphic organizer that had space to record the objects and processes in the model. Students characterized elements in the computer model as accurate, inaccurate, and missing. Assuming no prior knowledge of CT, the students were subsequently introduced to the notion that they could change the simulation through writing code "sentences." They then discussed ways to improve the computer model with most students deciding to add a kind of computer object, $MgSO_4$ molecules, not part of the original simulation.

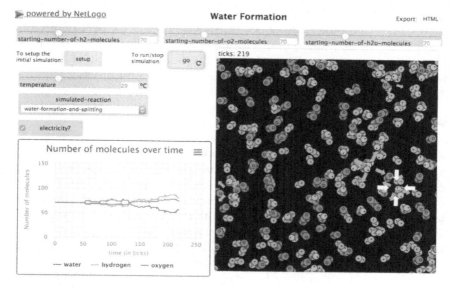

Figure 4.1. Screenshot of the NetLogo simulation of water formation and splitting.

Data Sources

Quantitative and qualitative data were obtained from multiple sources throughout the intervention. Two assessments, a computational attitude test (CAT) and a performance assessment of CT (PACT), were administered to students prior to the first CCT and again following the third CCT. The CAT consisted of Likert scale items that asked students to rate various attitudinal aspects with respect to computing, and the PACT consisted of tasks requiring CT, programming skills, and science content in varying degrees. In addition, digital images and original copies for student-generated artifacts were collected from all consenting students and analyzed qualitatively. These included models of the physical demonstration created by student groups and graphic organizers completed individually during the exploration of the simulation. Student activity in the simulation and its code was logged and analyzed for patterns. Throughout the intervention, video and audio recordings were captured for one class period for each teacher. One camera captured the teacher's instruction, while other cameras focused on group activities at the tables of students who had consented to participate in the research.

Video recordings were transcribed separately by members of the research team. As individual researchers produced transcriptions, they identified what they deemed as critical incidents (Flanagan, 1954) in the data, in this case those that highlighted teachers' pedagogical practices. These were then shared with other members of the research team for peer debriefing (Creswell & Miller, 2000) and joint qualitative analysis (Mercer, Dawes, Wegerif, & Sams, 2004). Discussions involved the comparison of incidents across teachers, to other sources of data, and to relevant literature.

Participants

Employing Yin's (2011) case study approach, we analyzed one teacher's approach to integrating CT into science instruction. While similar variation across domains was observed in other teachers' instruction, focusing on one teacher allowed for a richer discussion of qualitative data. Moreover, in comparison with the other three eighth grade teachers, learning gains were greatest amongst students in this teacher's classes. Figure 4.2 shows median grown across the four eighth grade teachers. For Casey's students, the median score grew 17 points from pretest to posttest, while for the other three teachers' students median scores grew by 11, 11, and 13 points. Therefore, we chose to focus analyses on this teacher's teaching in order to learn more about effective pedagogical strategies. To preserve anonymity and to personalize the discussion, we have assigned the pseudonym Casey. Casey was a White male eighth grade science teacher in an average SES. He had no prior experience teaching CT or programming. At the time of the intervention, he had eight years of teaching experience.

Since video data was collected in only one class period for each teacher, all analyses for this chapter draw from one videoed class period. Classes were 90-minute blocks taught every other day, and CCT1 spanned five class periods. A narrow focus on this classroom helped identify how varied pedagogical approaches functioned at the beginning of creating an integrated CT learning environment.

RESULTS

The following sections present data that highlight pedagogical patterns observed in Casey's classroom. The qualitative data from classroom interactions all occurred in the first of the three CCTs. The student assessment data discussed were gathered from pre/post performance assessment administered prior to CCT1 and following CCT3.

The following sections make the points that: (1) the teacher used whole class instruction to both setup and reflect upon exploratory learning; however, he also provided specific structures for students to think with, and (2) that in the CT instruction, where his domain expertise and pedagogical

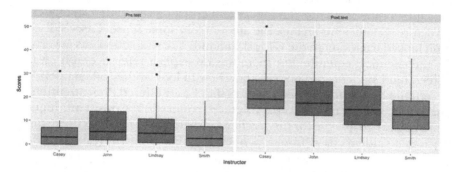

Figure 4.2.   Learning Gains on the Pre/postassessment across eighth grade teachers.

content knowledge (PCK) were more preliminary, the focus on developing students' concepts through discourse diminished.

Classroom Discourse and Learning

Most research grounded in constructivist theories of learning has taken a dichotomous approach to describing teacher-centered, direct as opposed to student-centered, exploratory pedagogy (Mascolo, 2009) describing them as conflicting teaching styles, differentiated by locus of control (Murphy, 2008). In previous publications, the authors have stated that it should not be an either-or-question; instruction should be wisely designed to optimize the benefits of both direct and exploratory modes of teaching (Martin & Herrera, 2007; Rivale, 2010; Rivale, Martin, & Diller, 2007). In practice, most teachers move back and forth between the two. In our observations, teachers guided the entire class to set up a group exploratory activity, probed individual students and student groups during exploration, gave just-in-time, whole-class, direct instruction and then followed up with a whole-class reflection on the exploratory activity. These interactions often involved use of questioning, some of which were compatible with the promotion of habits of mind associated with CT. Here we show how specific discursive moves affected the potential for learning of both science and CT.

Setup of guided exploration. During the intervention, Casey setup the exploration phase using whole-class instruction. He reminded students of the scientific core idea: "our phenomenon is the law of conservation of mass." He then wrote the balanced equation for the formation and splitting of water into H_2 and O_2 on the board and engaged the students in a discussion about the difference between the coefficients and subscripts in which he said, "I cannot change the small numbers [subscripts]; that changes the structure. Coefficients change the amount." In some sense, Casey's discussion looked irrelevant to the CT task at hand. The simulation was governed by the chemical equation but there was no explicit reference to it on the screen (and students did not yet know how to look at the code model).

Yet, the equation and the structures that Casey offered drew attention to the important science concepts to guide exploration. Consistent, higher, pedagogical goals were for the teacher's instruction to (1) focus the experience that students draw from a rich learning environment (Järvelä, 1995) and (2) served the crucial purpose of putting students in the position of being ready to learn (Murphy, 2008). In fact, Casey started off on a footing that emphasized *abstraction*, an important component of CT (Weintrop et al., 2015). Students were explicitly encouraged to think about the meaning of what the simulation depicted in relationship to their science knowledge.

During guided exploration. During student exploration of the simulation, Casey did not choose a hands-off approach. Instead, he visited students and probed for observations. Example 1 highlights an example of guided exploration facilitated through teacher-student interaction:

Example 1: Deciphering the Meaning of Arrows in the Simulation

Casey: Share a few observations.

Student: The white arrows mean formation.

Casey: Are you sure? How can you find out which arrows represent what? What option do you have in the controls to know exactly which ones represent what?

Student: Change it to water splitting

Casey: Yeah, and if water splitting is the only thing that is happening and you see the white arrows going towards each other, what do the white arrows going towards each other represent?

Student: Splitting

Casey could have simply told the student how to determine the meaning of the arrows using the simulation, an approach observed in other classes. Instead, Casey probed in a way that encouraged CT and required the students to do the higher order cognitive work of discovery. He kept his instructional focus on how the computer model could operate to provide an avenue to the answer. Key components of CT about models and representations are the modeler's ability to think in layers, using the objects and principles that are important at the level of representation chosen. Comments such as "if water splitting is the only thing that is happening and you see the white arrows going towards each other, what do the white arrows going towards each other represent?" bear a strong resemblance to the kinds of informal descriptions of objects and processes that computer scientists call *pseudo-code*. By emphasizing the meaning of the science rather than the procedures for exploration, not only did Casey's instruction complement the exploration rather than replicating but also helped structure student thought in a way that promoted CT. In this way, Casey guided the exploration in a way that encouraged more complex, model-based thinking from the student—the student did most of the higher order cognitive work.

Example 2 occurred not during the initial exploration of the model but rather as students worked to add a new object to their models. Casey used scientific principles to discuss molecule size with students and encouraged students to choose a size for the molecules that was compatible with its atomic mass.

Example 2: Atomic Mass of MgSO4

Casey: How many magnesium atoms [in MgSO4]?

Class: One

Casey: How many sulfur atoms?

Casey: How many sulfur atoms?

Casey: How many oxygen atoms?

Class: Four

Casey: *How many elements are present?*

Class: Three

Casey: How many total atoms?

Class: Six

Rather than beginning with the ultimate question of the number of atoms in $MgSO_4$, Casey used questions to walk students through each piece of information needed to determine how many atoms were in $MgSO_4$ as well as ensuring they understood that it is a compound of three separate elements. This method, referred to by Chin (2007) as "multi-pronged" questioning, employs a series of questions to promote deeper thinking on a concept. In fact, this exchange has the potential to evoke "multimodal thinking" (Chin, 2007), as he asked for a verbal response (verbal thinking) which was also linked to the mental image of the salt in the model (visual thinking) and rested on the symbols present in the chemical formula (symbolic thinking). Again, Casey's dissection of the concept of the $MgSO_4$ molecule into smaller components, leading students through each layer was more compatible with decompositional techniques such as used in CT. Such pedagogical exchanges can apprentice students in utilizing CT to dissect and reconceptualize their own understanding of the science.

Whole class guided reflection. Following exploration, Casey again positioned his teaching toward the entire class. His whole-class guided reflection encouraged students to participate throughout. Extending upon the tasks already visited on the graphic organizer during prior group work, Casey posed a number of related questions to the class that dug more deeply into the science modeled in the simulation. For example, he asked, "Can we say oxygen atoms?" and "What did the battery do for us in the simulation?" In this way, he extended the learning that occurred during exploration rather than simply reviewing it. In Example 3, Casey probed the students further by pushing them to think about an edge case. In computer science, an edge case is a problem that occurs only at an extreme maximum or minimum situation. Here, the minimum scenario highlighted the significance of the coefficients in the equation:

Example 3: Edge Case Thinking

Casey: If you only had one hydrogen molecule, would you ever get water to form?

Students: No

Casey: Why?

Student: Cause you need two

Casey: Yeah, this is our formation. Are single water molecules forming at a time?

Students: No

Casey: No, you are always getting?

Students: Two

Casey: That's our equation, ok. How many water molecules are forming at a time?

Students: Two

Casey encouraged students to verbalize their own thinking rather than verbalizing for them how they should think. Whatever merits or difficulties this presented as a program for science instruction, it is noteworthy how the kind of interaction he facilitated highlighted a kind of CT. In particular, inputs and outputs from the inferred processes become an interactive classroom focus.

Casey referred to both the law of conservation of mass and also the equation modeled at multiple points throughout the CCT. He did so in a way that continually grounded the CT exploration in the key science concepts modeled. Such observations evoked questions of whether students demonstrated evidence of learning gains in these areas.

Learning gains. A focal learning objective in CCT1 was understanding the law of conservation of mass, and a question on the pre/posttest asked students to name a scientific law to support the balance in the equation for water formation. Due to the consent process and student absences, we only have pre/posttest data from 11 of Casey's students. In Casey's class, only one student provided the law on the pretest, but six students did so on the posttest. It is possible that the prevalence of such scientific discourse during instruction helped the students in Casey's class to use the language of the discipline in the context of the posttest. Casey's approaches to classroom discourse, as seen in the examples presented, functioned to scaffold student learning, and this could have influenced learning gains as seen on the post assessment.

The most effective approaches to classroom discourse employ questions that require higher-order, cognitive thinking rather that those with a lower cognitive demand (such as questions that ask students to recall factual information) (Ong, Hart, & Chen, 2016; Redfield & Rousseau, 1981). Such effective approaches were observed in Casey's teaching. In addition, some approaches to content pedagogy may better align with integrating CT—a realization we came to after qualitative analysis of Casey's classroom discourse. The following section highlights the differing styles of teaching observed when the pedagogical focus shifted to CT. These observations led to the proposition that when integrating CT, teachers may benefit from more professional development, including guidance for classroom discourse, than what this project allowed.

Pedagogical Content Knowledge in the Integrated CT Classroom

Providing appropriate scaffolds implies that teachers are aware of students' learning needs (Murphy, 2008) and can tailor instruction accordingly. Doing this effectively requires sophisticated PCK (Rovegno, 1992; Shulman, 1986). In this section, we highlight a shift in a teacher's pedagogical approaches when the content domain changes, arguing for the role of teacher content knowledge in the integration of CT and programming. As the balance of science and CT/programming in the module shifted toward the latter, teachers relied more on the materials, thus limiting student involvement in the instruction.

As seen in the previous section, science content featured largely in the learning objectives and instruction on the first two days of CCT1. When focusing on science ideas, Casey layered periods of direct instruction with student exploration and evoked student commentary during discussions. On day 3, when the main objective shifted to learning programming skills, he shifted pedagogically to teacher-led approaches. The researchers had created instructional supports intended to guide teachers' implementation of programming instruction. Casey distributed copies of this teacher guide to each student and walked the class through the steps. He introduced the session with "I am going to walk you through like the first ten steps of this...I am going to guide you through this, ok?" In retrospect, it was a naïve expectation on the researchers' part that teachers could use the programming tools simply as a guide and employ more creative approaches while teaching it, as such creativity and differentiated approaches emerge from more intimate understanding of a discipline (Basista & Mathews, 2002). The tools provided to the teachers, therefore, were insufficient.

Programming instruction occurred in two separate facets of NetLogo. Students were first introduced to the *command center*—a window in the program where typing single-line commands could change the properties of objects in the simulation. Using the command center prior to introducing students to the program's full code familiarized them with some of the language's syntax and structure while demonstrating their power to change the simulation with simple statements. Casey either dictated the syntax of these commands for all students to type in unison or displayed sample commands on the Smartboard for them to copy. However, some students decoded the programming language themselves and were able to begin substituting other variables of the choice, thus assigning new properties to objects. Having observed several students' tendencies to creatively experiment in the code, we anticipate that training teachers how to encourage such endeavors could result in more accelerated learning among students.

After practicing making changes in the command center, Casey transitioned students to the program code where the goal was to implement a scientifically significant modification. The supplementary materials described adding Epsom salt, the catalyst used in the experiment, to the model. Researchers guided teachers through this process during the summer professional development workshop, and it was discussed that students could also add a battery or rust molecules to the simulation, but Casey chose to lead students through the processes for adding Epsom salt in the teacher's guide. We observed minimal discussion of why or how the computer language functions. Students did not engage in the planning step that asked them to make a table of the new object's properties, values, and a justification for each before changing the code. Instead, implementing the change involved copying several sections of existing code for the implementation of H_2O molecules and replacing "H_2O-molecules" with "$MgSO_4$-molecules."

This was a shift from how we had previously seen Casey teach science. Here, he utilized supplementary materials to directly walk students through each step. There were few opportunities for either student engagement or discussion. Unlike during the exploration of the simulation, students were not asked to generate or share any perceptions of the code, and there were no discussions of how the program code functioned to create the simulation with which they had been working. As previously cited, learning and practicing disciplinary language is crucial to learning that discipline. In fact, language learning is integral to learning how to practice in the programming domain (Duit & Treagust, 2012; Westgate & Hughes, 1997). Recall the earlier example of Casey explaining the role of the coefficients and the subscripts in a chemical equation. He knew that students needed to understand both the overall meaning of the equation as well as how each individual part functioned—both tied to scientific language. By contrast, the presentation of the programming language was more rote, allowing students to copy statements with little discussion of their interactive parts.

Furthermore, instances where Casey used scaffolding techniques in the programming context were likened to "debugging" (Klahr & Carver, 1988), as he simply helped students locate spelling or punctuation mistakes following an error message. There was little reflection on programming structure or functionality. Turkle and Papert (1992) argue that one of the benefits to learning of programming is that it encourages *epistemological pluralism*—multiple ways of knowing and doing. However, students were walked through the same exact coding steps, which left little room for creativity. This differed from earlier examples of teachers employing scaffolding or questions to invite students to think more deeply about the science.

Our observations support earlier arguments that integrating CT into science may require new content knowledge (programming/CT) as well as new PCK (CSTA & ISTE, 2011; National Research Council, 2011; Sendur et al., 2010; Yadav et al., 2014; Yadav, Zhou, Mayfield, Hambrusch, & Korb, 2011). It is one thing to know the discipline but another to know how to help others come to know that discipline, and we argue that to integrate CT effectively, teachers need not only to experience learning CT themselves but also to learn about how one learns CT. In spite of recent efforts to train teachers to teach computer science (CS for All Teachers, 2016), similar efforts do not exist to train core discipline teachers to integrate CT into their existing standards-based lessons.

While we saw Casey use tools and methods to engage students in deeper exploration in science learning, such use of methods or tools were missing in the computational tasks. Casey relied on the handout and enacted the steps one by one rather than scaffolding students' exploration (Wood, Bruner, & Ross, 1976) or probing them to think more deeply about the model and its functionality. As a result, the pedagogical balance swung closer to a whole-class walk-through rather than students exploring and discovering things in the code themselves. We attribute this both to time constraints but also to insufficient professional development and support for teachers in both their own understanding of programming and CT as well as how to integrate it throughout instruction. For example, as previously described, some students decided to change the sizes of all objects in the simulation to match their atomic masses. When working in the command center, students explored changing objects to shapes and sizes different from those presented by the teacher. Casey also offered some guidance toward accelerated activity in the code. He suggested to students who had implemented Epsom salt, "using your knowledge of how to add $MgSO_4$, see if you can make the appropriate changes to the code to add a battery." While possibilities for how students could modify the code were abundant, including advancing beyond the simple change outlined in the worksheet provided, teachers had not experienced implementing any other types of modifications to the code, and therefore, they did not possess the content knowledge necessary to guide students toward more advanced options. Still, our assessment data provided evidence of learning gains (Figure 4.2), and we see value in the continued integration of CT into science. Our work can point future researchers and administrators toward the development of more sophisticated approaches of supporting teachers in this innovative and challenging work.

DISCUSSION

Teachers are being asked not only to use technology in their teaching but also to integrate CT into core content areas. We looked at how this happened in the context of a project focused on integrating CT into science rather than on training teachers to do so.

Casey interacted with students in setting up activities, in looking at simulations and in recapping the simulations, in a way that aligned with concepts in CT. He pulled out issues of abstraction in modeling; he used pseudo-code like language; he structured discussion in a way that featured separation of concerns, and he highlighted edge case thinking. We did not provide any explicit guidance to participating teachers on how to teach in this way; we instead observed a natural connection between computational thinking and effective approaches to thinking about science, which this teacher employed. Initial video analyses of the other four teachers' pedagogy implied similar differences in their approaches to teaching science concepts versus their instruction of CT; however, focusing on one teacher illuminated these patterns more clearly. Our observations of Casey's style implied that other teachers integrating CT into their discipline could benefit from guidance toward a more synergistic approach to integrating the two domains.

While Casey's approach to teaching science aligned with CT, with minimal instruction in CT himself, he was not able, in this first offering of the intervention, to adapt these same purposes to the materials that focused more explicitly on computation, on programming and on changing the simulations. The problems with this should not be attributed to the teachers in this study. Teachers might similarly be limited if trying to integrate French into science with such little prior preparation. Such short-term instruction may enable the performance of simple tasks, but here we were asking teachers to integrate simulations and programs that were at the forefront of matters important to teaching eighth grade science: conservation of matter, chemical reactions, and other issues aligned with standards.

As a result, we recommend more research on how best to support teachers in this new endeavor in K–12 STEM education. Alongside the provision of new technology tools should come pedagogical guidelines, recommendations, and ongoing support that is tailored to the intervention. Teachers need opportunities to learn to program and to think computationally themselves before they can more effectively tailor instruction for their own population and context. This requires professional development opportunities beyond the provision of technology tools or written supports. In the same way that students must experience a domain and practice its language, teachers also should have opportunities to be initiated in the community of computer scientists.

However, while professional development can enhance teacher's integration of CT (Bean, Weese, Feldhausen, & Bell, 2015; Spradling, Linville, Rogers, & Clark, 2015), there are challenges. Unless funds are available to provide substitute teachers, schedules only allow for training outside of school hours and teachers may not want to give up their holidays or weekends without an incentive to do so. With limited time, it is not realistic to train core content teachers in the entire field of computer science. Creating integrated CT lesson plans is challenging for teachers (Bort & Brylow, 2013), and the question remains of what aspects of CS are most helpful to impart to teachers in short periods of training. Moreover, training should be differentiated (since teachers will come with varying levels of computing experiences) so one important aspect of professional development is to uncover any preconceptions about programming (Bean et al., 2015), an avenue worth deeper exploration.

Few studies have investigated best approaches to preparing teachers to teach CT (Blum & Cortina, 2007), and even fewer have focused on preparing non-CS teachers (Yadav et al., 2014). Our analyses of this teacher's classroom discourse revealed ways of communicating with students that are conducive to the development of computational thinking. Unfortunately, the short duration of this project did not allow for a second iteration of teacher professional development, but we are currently revising our materials and envisioning a redesign for teacher support that we hope to implement at other sites. Our aim is that future researchers and practitioners wishing to integrate CT can consider the subtle ways in which teachers' interactions with students can promote CT-aligned approaches to disciplinary thinking.

REFERENCES

Basista, B., & Mathews, S. (2002). Integrated science and mathematics professional development programs. *School Science and Mathematics*, *102*(7), 359–370. https://doi.org/10.1111/j.1949-8594.2002.tb18219.x

Bean, N., Weese, J., Feldhausen, R., & Bell, R. S. (2015). Starting from scratch: Developing a pre-service teacher training program in computational thinking. In *2015 IEEE Frontiers in Education Conference (FIE)* (pp. 1–8). https://doi.org/10.1109/FIE.2015.7344237

Blum, L., & Cortina, T. J. (2007). CS4HS: An outreach program for high school CS teachers. In *Proceedings of the 38th SIGCSE Technical Symposium on Computer Science Education* (pp. 19–23). New York, NY: ACM. https://doi.org/10.1145/1227310.1227320

Bonk, C. J., & King, K. (1998). *Searching for learner-centered, constructivist, and sociocultural components of collaborative educational learning tools | Electronic Collaborators | Taylor & Francis Group* (1st ed.). New York, NY: Routledge. Retrieved from https://www.taylorfrancis.com/books/e/9781136498565/chapters/10.4324%2F9780203053805-10

Bort, H., & Brylow, D. (2013). CS4Impact: Measuring computational thinking concepts present in cs4hs participant lesson plans. In *Proceeding of the 44th ACM Technical Symposium on Computer Science Education* (pp. 427–432). New York, NY: ACM. https://doi.org/10.1145/2445196.2445323

Bruner, J. (1996). *The culture of education*. Cambridge, MA: Harvard University Press.

Chin, C. (2007). Teacher questioning in science classrooms: Approaches that stimulate productive thinking. *Journal of Research in Science Teaching*, *44*(6), 815–843. https://doi.org/10.1002/tea.20171

Clemons, S. A. (2006). Constructivism pedagogy drives redevelopment of CAD course: A case study. *The Technology Teacher; Reston, 65*(5), 19–21.

Cobb, P., & Bowers, J. (1999). Cognitive and situated learning perspectives in theory and practice. *Educational Researcher, 28*(2), 4–15. https://doi.org/10.2307/1177185

Creswell, J. W., & Miller, D. L. (2000). Determining validity in qualitative inquiry. *Theory into Practice, 39*(3), 124–130. https://doi.org/10.1207/s15430421tip3903_2

CS for All Teachers. (2016). Computer science principles (CSP). NSF. Retrieved from https://csforallteachers.org/computer-science-principles

CSTA, & ISTE. (2011). *Computational thinking: Teacher resources* (2nd ed.). Retrieved from https://c.ymcdn.com/sites/www.csteachers.org/resource/resmgr/472.11CTTeacherResources_2ed.pdf

Cuny, J., Snyder, L., & Wing, J. M. (2010). Demystifying computational thinking for non-computer scientists. *Unpublished Manuscript in Progress, Referenced in https://www.cs.cmu.edu/link/research-notebook-computational-thinking-what-and-why*

De Freitas, S., & Neumann, T. (2009). The use of 'exploratory learning' for supporting immersive learning in virtual environments. *Computers & Education, 52*(2), 343–352. https://doi.org/10.1016/j.compedu.2008.09.010

Dickes, A. C., & Sengupta, P. (2013). Learning natural selection in 4th grade with multi-agent-based computational models. *Research in Science Education, 43*(3), 921–953. https://doi.org/10.1007/s11165-012-9293-2

diSessa, A. (2000). *Changing minds.* Cambridge, MA: The MIT Press.

Duit, R. H., & Treagust, D. F. (2012). Conceptual change: Still a powerful framework for improving the practice of science instruction. In *Issues and Challenges in Science Education Research* (pp. 43–54). Dordrecht, the Netherlands: Springer. https://doi.org/10.1007/978-94-007-3980-2_4

Flanagan, J. C. (1954). The critical incident technique. *Psychological Bulletin, 51*(4), 327. https://doi.org/10.1037/h0061470

Freeman, A., Adams Becker, S., & Cummins, M. (2017). *NMC/CoSN Horizon Report: 2017 K–12 Edition.* Austin, TX: The New Media Consortium. Retrieved from https://www.learntechlib.org/p/182003/

Goldstone, R. L., & Wilensky, U. (2008). Promoting transfer by grounding complex systems principles. *Journal of the Learning Sciences, 17*(4), 465–516. https://doi.org/10.1080/10508400802394898

Greeno, J. (1989). Situations, mental models, and generative knowledge. In D. Klahr & K. Kotovsky (Eds.), *Complex information processing: The impact of Herbert A. Simon* (pp. 285–317). Hillsdale, NJ: Psychology Press.

Grover, S., & Pea, R. (2018). Computational thinking: A competency whose time has come. In S. Sentance, S. Carsten, & E. Barendsen (Eds.), *Computer science education: Perspectives on teaching and learning in school* (pp. 19–38). London, England: Bloomsbury.

Herrmann-Abell, C., & E. DeBoer, G. (2011). Using distractor-driven standards-based multiple-choice assessments and Rasch modeling to investigate hierarchies of chemistry misconceptions and detect structural problems with individual items. *Chemistry Education Research and Practice*, *12*(2), 184–192. https://doi.org/10.1039/C1RP90023D

Israel, M., Pearson, J. N., Tapia, T., Wherfel, Q. M., & Reese, G. (2015). Supporting all learners in school-wide computational thinking: A cross-case qualitative analysis. *Computers & Education*, *82*, 263–279. https://doi.org/10.1016/j.compedu.2014.11.022

Jacobson, L. (2006). Wish list for schools. *Education Week*, *25*(23), 18.

Järvelä, S. (1995). The cognitive apprenticeship model in a technologically rich learning environment: Interpreting the learning interaction. *Learning and Instruction*, *5*(3), 237–259. https://doi.org/10.1016/0959-4752(95)00007-P

Kaput, J., & Schorr, R. (2008). Changing representational infrastructures changes most everything. The case of SimCalc, algebra, and calculus. In: *Research on technology and the teaching and learning of mathematics. Cases and perspectives* (Vol. 2, pp. 211–253). Charlotte, NC: NCTM and Information Age.

Klahr, D., & Carver, S. M. (1988). Cognitive objectives in a LOGO debugging curriculum: Instruction, learning, and transfer. *Cognitive Psychology*, *20*(3), 362–404. https://doi.org/10.1016/0010-0285(88)90004-7

Lave, J., & Wenger, E. (1991). *Situated learning: Legitimate peripheral participation*. Cambridge, UK: Cambridge University Press.

Lee, O., Eichinger, D. C., Anderson, C. W., Berkheimer, G. D., & Blakeslee, T. D. (1993). Changing middle school students' conceptions of matter and molecules. *Journal of Research in Science Teaching*, *30*(3), 249–270. https://doi.org/10.1002/tea.3660300304

Lye, S. Y., & Koh, J. H. L. (2014). Review on teaching and learning of computational thinking through programming: What is next for K–12? *Computers in Human Behavior*, *41*, 51–61. https://doi.org/10.1016/j.chb.2014.09.012

Martin, T., & Herrera, T. (2007). *Mathematics teaching today: Improving practice, improving student learning*. Reston, VA: National Council of Teachers of Mathematics.

Mascolo, M. F. (2009). Beyond student-centered and teacher-centered pedagogy: Teaching and learning as guided participation. *Pedagogy and the Human Sciences*, *1*, 3–27.

Mercer, N., Dawes, L., Wegerif, R., & Sams, C. (2004). Reasoning as a scientist: Ways of helping children to use language to learn science. *British Educational Research Journal*, *30*(3), 359–377. https:// doi.org/10.1080/0141192041000 1689689

Murphy, P. (1996). Defining pedagogy. In P. Murphy & C. F. Gipps (Eds.), *Equity in the classroom: Towards effective pedagogy for girls and boys*. London, England: Falmer Press.

Murphy, P. (2008). Gender and subject cultures in practice. In P. Murphy & K. Hall (Eds.), *Learning and practice: Agency and identities* (pp. 161–172). London, England: SAGE.

National Research Council. (2010). *Report of a workshop on the scope and nature of computational thinking*. Washington, DC: National Academies Press.

National Research Council. (2011). *Report of a workshop on the pedagogical aspects of computational thinking*. Washington, DC: National Academies Press.

NGSS Lead States. (2003). *Next generation science standards: For states, by states.* Washington, DC: National Academies Press.

Ong, K. K. A., Hart, C. E., & Chen, P. K. (2016). Promoting higher-order thinking through teacher questioning: A case study of a Singapore science classroom. *New Waves, 19*(1), 1–19.

Papert, S. (1980). *Mindstorms: Children, computers, and powerful ideas.* New York, NY: Basic Books.

Puttick, G., Drayton, B., & Karp, J. (2015). Digital curriculum in the classroom: Authority, control, and teacher role. *International Journal of Emerging Technologies in Learning (IJET), 10*(6), 11–20. https://doi.org/10.3991/ijet.v10i6.4825

Qualls, J. A., & Sherrell, L. B. (2010). Why computational thinking should be integrated into the curriculum. *The Journal of Computing Sciences in Colleges, 25*(5), 66–71.

Redfield, D. L., & Rousseau, E. W. (1981). A meta-analysis of experimental research on teacher questioning behavior. *Review of Educational Research, 51*(2), 237–245. https://doi.org/10.3102/00346543051002237

Reed, D. A., Bajcsy, R., Fernandez, M. A., Griffiths, J.-M., Mott, R. D., Dongarra, J., … Ponick, T. L. (2005). *Computational science: Ensuring America's competitiveness.* Arlington, VA: President's Information Technology Advisory Committee. Retrieved from http://www.dtic.mil/docs/citations/ADA462840

Rivale, S., Martin, T., & Diller, K. R. (2007, June). Comparison of Student Learning in Challenge-based and Traditional Instruction in Biomedical Engineering. *Proceedings, American Society of Engineering Education, Honolulu, HI.*

Rivale, S. (2010). *An expert study in heat transfer* (Doctoral dissertation). University of Texas at Austin). Retrieved from https://repositories.lib.utexas.edu/bitstream/handle/2152/23491/RIVALE-DISSERTATION-2010.pdf?sequence=1

Rogers, E. (1995) *Adoption of innovation.* New York, NY: Free Press.

Rovegno, I. C. (1992). Learning to teach in a field-based methods course: The development of pedagogical content knowledge. *Teaching and Teacher Education, 8*(1), 69–82. https://doi.org/10.1016/0742-051X(92)90041-Z

Schwarz, C. V., Reiser, B. J., Davis, E. A., Kenyon, L., Achér, A., Fortus, D., … Krajcik, J. (2009). Developing a learning progression for scientific modeling: Making scientific modeling accessible and meaningful for learners. *Journal of Research in Science Teaching, 46*(6), 632–654. https:/doi.org/10.1002/tea.20311

Sendur, G., Toprak, M., & Pekmez, E. S. (2010, Month?). *Analyzing of students' misconceptions about chemical equilibrium.* Presented at the International Conference on New Trends in Education and Their Implications. LOCATION?

Sengupta, P., Kinnebrew, J. S., Basu, S., Biswas, G., & Clark, D. (2013). Integrating computational thinking with K–12 science education using agent-based computation: A theoretical framework. *Education and Information Technologies, 18*(2), 351–380. https://doi.org/10.1007/s10639-012-9240-x

Settle, A., Franke, B., Hansen, R., Spaltro, F., Jurisson, C., Rennert-May, C., & Wildeman, B. (2012). Infusing computational thinking into the middle- and high-school curriculum. In *Proceedings of the 17th ACM Annual Conference on*

Innovation and Technology in Computer Science Education (pp. 22–27). New York, NY: ACM. https://doi.org/10.1145/2325296.2325306

Shulman, L. S. (1986). Those who understand: Knowledge growth in teaching. *Educational Researcher, 15*, 4–14. https://doi.org/10.3102/0013189X015002004

Spradling, C., Linville, D., Rogers, M. P., & Clark, J. (2015). Are MOOCs an appropriate pedagogy for training K-12 teachers computer science concepts? *Journal of Computing Sciences in Colleges, 30*(5), 115–125.

Stavy, R. (1991). Children's ideas about matter. *School Science and Mathematics, 91*(6), 240–244. https://doi.org/10.1111/j.1949-8594.1991.tb12090.x

Texas Education Agency. (2013). *Texas essential knowledge and skills for science, subchapter bNGSS. Middle School.* TEA. Retrieved from http://ritter.tea.state.tx.us/rules/tac/chapter112/ch112b.html

Turkle, S., & Papert, S. (1992). Epistemological pluralism and the revaluation of the concrete. *Journal of Mathematical Behavior, 11*(1), 3–33.

van Oers, B. (2000). The appropriation of mathematical symbols. In *Symbolizing and Communicating in Mathematics Classrooms: Perspectives on Discourse, Tools and Instructional Design* (pp. 133–176). Mahwah, NJ: Lawrence Erlbaum Associates.

Vygotsky, L. S. (1978). *Mind in society: The development of higher psychological processes.* Cambridge, MA: Harvard University Press.

Weintrop, D., Beheshti, E., Horn, M., Orton, K., Jona, K., Trouille, L., & Wilensky, U. (2015). Defining computational thinking for mathematics and science classrooms. *Journal of Science Education and Technology, 25*(1), 127–147. https://doi.org/10.1007/s10956-015-9581-5

Wenger, E. (1998). *Communities of practice: Learning, meanings, and identity.* Cambridge, England: Cambridge University Press. Retrieved from http://www.chris-kimble.com/Courses/mis/Communities_of_Practice.html

Westgate, D., & Hughes, M. (1997). Identifying "quality" in classroom talk: An enduring research task. *Language and Education, 11*(2), 125–139. https://doi.org/10.1080/09500789708666723

Wilensky, U. (1999). *NetLogo.* Evanston, IL: Center for Connected Learning and Computer-Based Modeling, Northwestern University. Retrieved from http://ccl.northwestern.edu/netlogo/

Wilensky, U., & Reisman, K. (2006). Thinking like a wolf, a sheep, or a firefly: Learning biology through constructing and testing computational theories— an embodied modeling approach. *Cognition and Instruction, 24*(2), 171–209. https://doi.org/10.1207/s1532690xci2402_1

Wilensky, U., & Stroup, W. (1999). Learning through participatory simulations: Network-based design for systems learning in classrooms. In *Proceedings of the 1999 Conference on Computer Support for Collaborative Learning.* Palo Alto, CA: International Society of the Learning Sciences. Retrieved from http://dl.acm.org/citation.cfm?id=1150240.1150320

Wood, D., Bruner, J. S., & Ross, G. (1976). The role of tutoring in problem solving. *Journal of Child Psychology and Psychiatry, 17*(2), 89–100. https://doi.org/10.1111/j.1469-7610.1976.tb00381.x

Yadav, A., Mayfield, C., Zhou, N., Hambrusch, S., & Korb, J. T. (2014). Computational thinking in elementary and secondary teacher education. *Transactions on Computing Education*, *14*(1), 5:1–5:16. https://doi.org/10.1145/2576872

Yadav, A., Zhou, N., Mayfield, C., Hambrusch, S., & Korb, J. T. (2011). Introducing computational thinking in education courses. In *Proceedings of the 42nd ACM Technical Symposium on Computer Science Education* (pp. 465–470). New York, NY: ACM. https://doi.org/10.1145/1953163.1953297

Yin, R. K. (2011). *Applications of case study research*. London, England: SAGE.

CHAPTER 5

DEVELOPING ELEMENTARY STUDENTS' PROBLEM SOLVING, CRITICAL THINKING, CREATIVITY, AND COLLABORATION THROUGH A UNIVERSITY-SCHOOL PARTNERSHIP

Nancy Streim, Susan Lowes, Elizabeth Herbert-Wasson, Yan Carlos Colón, Lalitha Vasudevan, Jung-Hyun Ahn, and Woonhee Sung
Teachers College Columbia University

This chapter explores a university-school partnership that leverages technology for *advancing a culture of innovation* in schools using STEAM as a vehicle (Freeman, Adams Becker, Cummins, Davis, & Hall Giesinger, 2017). The authors describe experiences teaching technology-mediated STEAM enrichment in a New York City elementary school. The chapter presents three cases of university led courses: a digital literacy curriculum focused on expanding students' agency for their learning; 2 year-long music composition courses

Integrating Digital Technology in Education:
School–University–Community Collaboration, pp. 111–134
Copyright © 2019 by Information Age Publishing
All rights of reproduction in any form reserved.

that use technology to foster collaborative music making; and a robotics club that also serves as a research project to examine the impact of embodied learning for teaching science and math concepts. We share accomplishments and challenges that speak to the practical realities of teaching STEAM through a school-university partnership. We also relate our experiences to three themes raised in the recent NMC/CoSN Horizon Report on the state of technology integration in schools (Freeman et al., 2017). We conclude that university-school partnerships can directly engage elementary school students in learning experiences that strengthen skills in problem solving, critical thinking, creativity, and collaboration, although state accountability systems can inhibit transfer of related pedagogical principles to the core curriculum. If our experiences are representative, advancing a culture of innovation through STEAM is likely to remain the domain of non-tested subject areas for the time being. However, our cases also illustrate the benefit of giving future teachers, educational researchers, and school leaders opportunities to develop their understanding about the practicalities of integrating technology into teaching and learning.

TEACHING COMPUTATIONAL THINKING THROUGH STEAM

Integration of technology into K–12 education is advancing at a feverish pace. Freeman et al. (2017) identified six short-, medium-, and long-term trends that are transforming schools into digitally rich learning environments. Of particular interest to us is the potential of university-school partnerships to support the long-term trend cited by Freeman et al. toward leveraging technology for *advancing a culture of innovation* in schools with STEAM as a vehicle for fostering a culture, marked by project-based learning that features entrepreneurship, collaboration, and creativity. In this chapter, we focus on the authors' experiences introducing technology-mediated STEAM-based enrichment experiences to develop elementary students' computational thinking skills in a New York City elementary school.

The International Society for Technology in Education (ISTE) defined computational thinking as a problem-solving process that includes "formulating problems in a way that enables us to use a computer and other tools to help solve them" (ISTE, 2011). Since the term is often used in reference to computer science education, others have substituted *complex thinking, problem based, inquiry based,* or *real world thinking* to describe these and similar skills (Freeman et al., 2017). We prefer the definition of computational thinking offered by the National Science Foundation (n.d.), which described it as an approach that enables students to "develop skills and competencies in problem solving, critical thinking, creativity, and collaboration." Computational thinking, conceived this way, is relevant to STEAM learning as it marries the stepwise, sequential aspects of problem solving as found in computer science with the less rule-bound and subjective possibilities of artistic expression (Sousa & Pilecki, 2013).

STEAM is also a fertile context for students to practice computational thinking through pursuit of high interest projects that encourage innovation. As noted by STEAM champion (2013) at a briefing for the Congressional STEAM Caucus: "Innovation depends on the problem solving, risk taking, and creativity that are natural to the way artists and designers think. Art and science—once inextricably linked—are better together than apart."

Most schools today are equipped with hardware and connectivity, and the rapid introduction of new software applications is expanding the tools for engaging students in computational thinking. However, to what extent is the presence of technology a signal that schools are taking advantage of its potential for transforming learning? If not, what is required to make the transition from using technology to accomplish traditional school tasks to designing student-centered approaches that harness technology to promote a culture of innovation? And what is the role of higher education in helping schools realize the potentialities of digital learning?

IMPORTANCE OF PREPARING STUDENTS WITH DIGITAL SKILLS AND COMPETENCIES

Freeman et al. (2017) argued:

> Developing skills that enable learners to use computers to gather data, break it down into smaller parts, and analyze patterns will be an increasing necessity to succeed in our digital world. While coding is one aspect of this idea, even those not pursuing computer science jobs will need these skills to work with their future colleagues. (p. 4).

Though not all career preparation need be characterized in relation to computer science, as digital technologies come to dominate our professional and personal lives, it is critical for schools to prepare students to be full participants as careers across most disciplines already require digital skills for seeking and analyzing information; collaborating and communicating with others; and presenting or reporting information. Arguably, most schools are exposing students to digital learning, information seeking, and presentation skills, yet a missing piece appears to be transforming curriculum and pedagogy to incorporate innovative learning strategies that introduce these skills in more open-ended and student driven settings that promote what Freeman et al. (2017) called "agile and collaborative mindsets" (p. 12).

In the sections that follow, we introduce a university-school partnership focused on STEAM enrichment programming in which we sought to better understand implementation of problem solving, critical thinking, and creative and collaborative aspects of computational thinking in learning contexts characterized by innovation and student agency.

UNIVERSITY-SCHOOL PARTNERSHIP

In this chapter, we present three cases that describe our experiences teaching STEAM enrichment courses in robotics, digital literacy, and music composition at the Teachers College Community School (TCCS), where we work with students and staff in a unique university-school partnership. We explore the processes of exposing elementary students to computational thinking that is meant to foster authentic learning and capitalize on experimentation, creativity, and collaboration.

TCCS is a pre-K–8 public school in the West Harlem neighborhood of New York that was established in 2011 through a formal agreement between Teachers College (TC) and the New York City Department of Education (NYCDOE). It is a regular public city school, managed by the NYCDOE. The establishment of TCCS is a signature initiative of TC, which has made a long-term commitment to guide and support the school. The authors are faculty, staff, and graduate students at TC, and the first author is responsible for the stewardship of TCCS as a demonstration model of a university-assisted public school (Streim, 2016).

School Characteristics

TCCS is a "choice" elementary school, which means that students are admitted by lottery at three entry points: prekindergarten, kindergarten, and Grade 6. There are no admissions criteria although there is a selection preference for children living in the Harlem communities located near the TC campus. In the 2017–18 school year, the student body numbered 290 in grades prekindergarten through Grade 6, and it is expected to reach 470 students when the school completes its trajectory to eighth grade in the 2019–20 school year. The school attracts an ethnically diverse student body with 40% Latinx, 36% African American, and 15% Caucasian students. Sixteen percent receive special education services. Nearly half the students qualify for free or reduced-price lunch. TCCS occupies a former parochial school building located one-half mile from the TC campus. The traditional learning spaces in the building are sometimes constraining, but the school environment is welcoming, intimate, and homey. There is no expectation that the school will exclusively hire teachers who are graduates of TC although, currently, 25% of the teachers at TCCS are TC alumni.

School Context

As stated on its website:

The TCCS mission is to establish a dynamic learning environment that nourishes children's academic, social, emotional, and physical well-being. TCCS strives to provide a child-centered environment that will inspire and challenge all of our students to become independent thinkers, problem solvers, and life-long learners, and to work as a collaborative unit of parents, faculty, and staff to ensure that our students perform their optimal best. (Teachers College Community School, n.d.)

Still in its development stage, TCCS is already a thriving learning community and boasts a high degree of teacher leadership and parent engagement. While other elementary schools in the area are losing students to charter schools, 600 students applied for 118 seats at TCCS for the upcoming school year. The school is obviously popular with families; nevertheless, maintaining the positive learning culture while continually improving performance on external academic measures is a high priority for the school.

Three special features of the school are relevant to teaching computational thinking through a university-school partnership. First, many TC students are active as instructors, tutors, interns, classroom consultants, and researchers. Second, the school has a rich music program sponsored by TC. With TC graduate students as their instructors, third grade students learn to play the violin, fourth graders participate in choir, and fifth and sixth graders learn digital composition. Third, the school prides itself on the availability of technology. Through grants and strategic allocation of its budget, the school administration has acquired desktops, laptops, iPads, and smartboards to serve every classroom. Teachers regularly incorporate learning software into instruction. The school has added coding and robotics into the upper elementary and middle school curriculum. Students use Google classroom to track work and collaborate on group projects; they create multimedia presentations, conduct online research, and use software to develop fluency in reading and math.

Despite the availability and utilization of technology, there is limited evidence that technology is being optimized to teach computational thinking. A TC doctoral student conducted a visioning exercise with staff in 2016 and found that most teachers are highly interested in how they might use technology more effectively but do not have models for what progressive, student-centered, technology-mediated learning looks like in the class-room. Indeed, the principal noted that incorporating technology is at an early stage. It is a "catalyst to get the work done, not to push kids' thinking" (M. Verdiner, personal communication, May 22, 2018).

THREE CASES

In order to contribute to the school's STEAM curriculum, engage students in computational thinking, and model a learning culture of innovation for students and teachers, TC faculty and doctoral students designed a number of STEAM experiences for the children at TCCS. We had three goals: to enrich the curriculum for students, to model how teachers can develop similar experiences in the core subject areas, and to study the impact of computational thinking on learning. The cases presented here represent three examples: a digital literacy curriculum focused on expanding stu-dents' agency for their learning; 2 year-long music composition courses that used technology to foster collaborative music making, and a robotics club that also served as a research project to examine embodied learning as a tool for teaching science and math concepts.

All the instructors were masters or doctoral students working under the guidance of a faculty mentor. They each received compensation for teach-ing, in the form of a fellowship, stipend, or part-time student employment. And all were using the experiences to also pursue research questions or conduct pilot studies related to their academic interests. The students share their experiences in their own voices below.

Case 1: Creativity and Collaboration Through Digital Music Composition (Yan Carlos Colón)

As a doctoral candidate in music education at TC, I have been inter-ested in the possibilities of incorporating music with STEAM learning. I have also been deeply involved in teaching music at TCCS. When I was approached to propose a middle grade music sequence at the school, I designed two courses in which TCCS fifth and sixth grade students merge art and technology as they create music collaboratively. My approach was based on Kanellopoulos (2012) who explored collaborative music-making, through improvisation and composition, and empha-sizes that collective music-making can be part of a process by which we cultivate democracy. In developing these courses, I was concerned with encouraging collaboration, community building, and opportunities for democracy in practice. The result was a 2-year music experience in which students, in the context of digital technology, compose film music by themselves and/or with their peers.

Music composition is a complex activity, and students are constantly engaging in problem solving and critical thinking. Democracy is evident when every voice is taken into consideration and when diversity is an advantage instead of a breaking point. In our class, students created the music that they wanted, using the sounds that they wanted. Having these choices helped them develop ownership over their own learning processes and musical evolution.

Fifth grade music composition. I have taught this course for 2 years. Two classes of 25 students met once a week in the school's library. The principal had a teacher stay with the students for the period when I taught, with the idea that classroom teachers can learn about their students from watching them in a project-based learning context and, thus, gain insights for transferring project-based learning principles to other curricular areas.

I divided the music composition course into two parts. During the first semester, the students worked individually as they learned to set up their work stations, which were composed of a laptop, a midi-controller keyboard, and a set of headphones. Students became acquainted with the software by completing a series of projects that explored the software as a tool for digital creation. Simultaneously they worked on compositional exercises that helped them understand techniques such as variation and organic development of musical themes and ideas. By the end of the first semester, the students became aware of their own strengths and weaknesses with the software and with composing in general.

The aim of the second semester was to collaboratively create a piece of music. Working in groups of three to five students, they experienced composing, producing, and recording their music in a framework where process and product were equally important. Every group had one laptop and the students assigned themselves roles based on their assessment of what would benefit the group's work flow. I provided support, feedback, advice, and technical assistance throughout the process, which culminated in a song or track. Every track created over the two years included multiple digital instruments, lyrics, and vocals. The music created by the students was diverse, not only in sounds, but in instrumentation, texture, harmony, and form. In the end, the product was not only evidence of their evolution as composers and musicians but was a testament of how they were able, despite their differences, to create, collaborate, and compose together.

Sixth grade film scoring. Having exposed students to collaborative compositional and production processes in fifth grade, the aim of this class was for students to create music for short clips of film. One class of 20 students met weekly in the school's library during the 2017–18 school year. In film scoring, the music is at the service of the image and must relate to both the storyline and other artistic decisions made by the film's creator. The challenge for the students is to find a way to be creative within these parameters. Film scoring also represents a collaboration across the art forms of music and film, in which students navigate the challenges of adding themselves to the fixed medium of an existing film. Students were allowed to use samples and sound effects from GarageBand's library to complement original compositions.

The library setting gave us space to move around. The students sat at tables, sometimes in groups, sometimes by themselves. Some preferred the rug, and others gravitated to the bean bag chairs. Even when students were working on individual projects, they sat close to each other and they shared ideas, sounds, and strategies. They showed each other their process and when they felt they had accomplished something outstanding, they wanted to share it with everyone. Yes, it got loud and messy sometimes, but in this fluid experience, in their very own way, they supported and influenced each other.

Our first project this year was to musicalize the classic "Walt Disney Pictures" opening clip that appears at the start of every Disney movie. The students worked by themselves with the goal of replacing the existing music with their own version. Their work would be later shared and discussed with the rest of the class.

In the short but rich opening clip, the students first identified details such as a smoking train that appears in the distant background, the flag-pole coming into view at the top of the castle, a fireworks display, and the moment when the typography for "Walt Disney Pictures" comes into view. Responding to the images, one student scored a siren to make the train come to life, another introduced a sample of a hip-hop beat when the flagpole appeared. One student used percussion instruments to simulate the sound of fireworks, and another synchronized the appearance of the typography with a sudden stop in the music, which resulted in a dramatic presentation of the name of the production house.

The second project involved more complex film clips. One of the favorites among the students was an excerpt from Disney's *Tron: Legacy* with original music by Daft Punk. The students had to synchronize action and sound and also accommodate changes of mood, conflicts, and other emotions. Students that selected this film clip had to score a fast-paced police motorcycle chase scene followed by a contemplative one of a security guard who fails to react when the motorcycle's image travels across his security monitors.

Challenges and lessons learned. When I was designing the courses, I was concerned that the technological aspects would be difficult for the students. After all, I have been working with music technology for more than 15 years and it took me a long time to become proficient. I quickly discovered that technology was not a challenge for the students at all. Providing them with digital tools placed them at the appropriate starting line for composing. They were able to navigate new technological experiences with a capacity to quickly adapt and discover new ways in which technology can serve their purposes.

To my surprise, the biggest challenge was facilitating the social aspect of creating music together. Giving the students freedom to make most decisions about their work required them to assume a high degree of autonomy and responsibility (Freire, 1970). Open-ended assignments were difficult for some students who became frustrated that expectations were not clear. Other students initially lacked confidence to have their ideas discussed by a group because of the risk that their contributions would be dismissed or transformed. Instead, many students would advocate for their own views as if the exercise were a competition and then feel rejected if their ideas did not win out. The experience of rejection engendered resentment and some sought to work alone.

At these moments, I no longer felt like a music composition teacher, but had to become a mediator and a counselor. I made it a point to have a pop-up discussion with a group in conflict. I guided the students to reflect on their actions and how they were affecting the group. We then processed strategies to move forward in a way that was favorable for all. For example, one of the fifth grade groups had trouble reaching consensus on a direction for their project. Each student had a different perspective and became frustrated as other groups made progress while theirs remained stalled. In our discussions, I addressed the subject of *difference* and how it can either be a divisive agent or an advantage because having multiple ideas provides more possibilities. This group struggled to arrive at a breakthrough moment but they succeeded and their project moved quickly. They used their different perspectives to identify the multiple talents and skills in the group and assigned roles accordingly. They accommodated differences by breaking into smaller groups to accomplish specific tasks and then reunited to assemble the parts into a finished whole. While some of the students worked on lyrics on a piece of paper, another subgroup worked on the beat in the computer or just found that cool sound they were seeking for their melody. With patience and compromise they accomplished the goal of working together to complete their project.

After taking the time to work through difficulties, beautiful breakthroughs can happen. I have seen students become more empathic and help each other—from carrying a heavy midi-controller to explaining how to edit and quantize midi. I have also seen students scaffold their understanding of concepts across contexts. For example, one student in the film scoring class, who is also a cellist in the orchestra, used a combination of original and sampled orchestral music to create a texture that achieved a dramatic effect in the police motorcycle chase scene. She represented the changes in mood and movement using a musical form as she understood it through her experience in orchestra. She probably would not verbalize her compositional thoughts as I have, but I could recognize clear and conscious techniques in the choices she made and she had no trouble expressing them with the technological tools available to her.

At a more practical level, I also confronted challenges related to working conditions in schools. Even though the music program had strong support from parents and administration, securing optimal spaces for teaching could be a challenge. Since TCCS added one new grade each year, we eventually lost our music classroom. The music teachers adapted to moving around the school as necessary—to the library, the stage, or the large multi-purpose room. Consistency in scheduling was another challenge. For example, when student testing occurred several times a year, I might suddenly lose access to the classroom where the music technology was set up. Even if my students were not taking tests, classroom changes prevented us from accessing the materials we normally used, forcing the students to interrupt their workflow. Since my classes met for only 45 minutes every week, schedule changes due to assemblies, field trips, and testing could be a significant setback in meeting course goals. On the other hand, the students adapted by becoming very aware of the preciousness of class time. They came to the room ready to work, and they set up and took down their own work stations. By understanding how fragile schedules can be, the students learned to work efficiently, making every second count.

What I have learned from the experience of infusing music and technology suggests that technology is not the challenge. The challenge is how we encourage progressive teaching in our classes through a curriculum that develops creativity and collaboration. Probing the relationship between arts and technology is a fertile context for building these skills.

Case 2: Enhancing Digital Literacy Skills Through Student-Centered Learning (Elizabeth Herbert-Wasson and Lalitha Vasudevan)

This case addresses our experiences leading a year-long series of workshops whose primary goal was to help students develop their proficiency when working with digital technologies by building on their existing interests and skills. Other objectives included establishing a sense of collegiality among students, and between students and their teachers; encouraging exploration of unfamiliar tools and software; and designing, planning, and implementing an independent or group project from start to finish. Central to our approach was an embrace of the unexpected, pedagogical flexibility, ample opportunities to play with technologies and media, and multiple modes of participation and communication.

In our practice, we draw on the concept of agency as described by Lewis and Moje (2003): "Agency ... can be thought of as the *strategic* making and remaking of selves; identities; activities; relationships; cultural tools and resources; histories" (p. 1985). The course we describe here builds on research that looks at the artifacts students produce during the project development process rather than at summative assessment data and considers student achievement from a more generative perspective of how it may lead to desired educational outcomes and improved teacher-student relationships. For instance, Clark's (2011) work showed how young children make meaning visible through the creation of multimodal maps. Through their work in youth radio, Chávez and Soep (2005) noted the relational shifts that occur between adults and children when they work collaboratively rather than hierarchically. Sullivan's (2011) study of how individuals learn from the creative work of others also supported the use of a more participatory approach to pedagogy within a public school classroom.

The collaboration with TCCS was based on prior work in a New York City after-school program. Having developed our approach in an out-of-school setting, we wanted to see if the pedagogy would transfer into an in-school classroom environment. Following a short module that was taught to fourth graders in the spring of 2017, author Elizabeth Herbert-Wasson and another graduate student worked with three different classes of fifth and sixth graders during the 2017–2018 school year. Within these classes, students worked in self-selected groups of one to eight students. The workshops took place during one of the regularly scheduled theater class periods each week. This model was chosen at the beginning of the year because it allowed for collaboration between the project facilitators and a single teacher for all classes and worked well for scheduling purposes.

We introduced the project with the broad prompt of, "We're going to make something that has to do with both theater and technology. We have computers and cameras available, as well as other equipment if requested. What would you want to make?" Students then spent the next several weeks discussing possible topics with classmates, testing out different possible groupings, and experimenting with equipment. We had students making games in Scratch and Gamestar Mechanic, videos with digital cameras and iMovie, audio compilations using GarageBand and SoundCloud, and websites through Wix. Many students moved between groups well into the year and several groups transformed their ideas into artifacts very different from what they had discussed at the outset of the project.

As part of the path to creating the final artifacts, we encouraged iterative, rather than linear design and collaboration processes, and recognized that not all prototypes would make it into the final artifacts. This led to often circuitous conversations that pushed students to resolve disagreements with group members, reason through challenges, and identify solutions to problems themselves.

Similar to the experience noted by Colon-Leon in the first case, our students quickly embraced the technology. They were deeply engaged in the creation process and designing and managing their own projects, including getting to work as soon as our sessions began. The facilitator's role was largely to help identify digital tools, apps, and platforms that supported the students' design interests and then to point students in the direction of resources that might help them further develop their mastery of the tools.

While their passion and excitement for the projects was obvious in the artifacts they were developing, this work was not quiet. Students were frequently in the hallways looking for places to film. Group members talked over each other trying to share ideas. Students sat on desks to get a better view of a classmate's computer screen. Sometimes a student called across the room to get someone else's attention, and entire groups occasionally collapsed into a ball of laughter if they struck upon a particularly amusing idea.

Final artifacts began to emerge including a multi-level platformer video game containing theater concepts, a series of film shorts curated on a student-made website, a filmed walkthrough of a virtual theater created with Minecraft, a musical piece written by students and recorded in GarageBand, and a movie combining both live action shots and digitally created graphics.

Challenges and lessons learned. Students' enthusiasm during our workshops was usually joyous but also very much unrestrained. This type of classroom environment can be jarring if a teacher expects to maintain control over what students are doing and how they are carrying out their tasks. Traditional classroom management norms are frequently abandoned and students are given the space to learn how to navigate group dynamics and the project development process themselves. As a result, it can be difficult to find a balance between helping students unleash their creative potential and ensuring a space that is compatible with the culture and practices of the broader school community. Such a pedagogy requires flexibility and playfulness. Because of the generative nature of this project, developing a sense of collegiality between adults and students is at the center of this work. The boundaries between the rigid roles of teacher and student must be more permeable, allowing teachers to also be students and students to also be teachers. Any teacher who is considering embarking on a project like this should reflect about whether he or she is prepared for what may seem like occasional, albeit glorious, chaos.

We also learned quite a bit from logistical hurdles that challenged our pedagogical expectations. At certain times of the year, our weekly sessions began turning into monthly sessions because of holidays, snow days, field trips, and the demands of standardized testing. The specific laptops available to us sometimes differed from week to week and many of the student projects were in locally stored files. Since students did not have access to a cloud-based file management system, we began collecting student work on a flash drive which consumed a large percentage of their work time. Also, some students wanted to access programs that were blocked by the school's Internet filter, or they wanted to use music and other files on their phones, whose use in class is against school policy. Also, because of equipment scheduling issues during one period of the day, we shared a laptop cart with another class for much of the year. As a result, what were intended to be individual projects became group projects. As Colon discussed in relation to the music classes, group projects require the development of communication and (often) conflict resolution skills which we had not anticipated in our original project.

In moving forward with this type of student-driven course, it will be extremely important to have the logistics organized from the outset. Also, as schools integrate more creative technology-based learning, they may need to revisit their technology plans and procedures. School administrators might consider questions such as: Does the laptop cart model still work for us? How best can we share resources? At what times might we need dedicated equipment? What other needs are emerging when we try to integrate more technology into our classes? How can digital tools help to reveal and support what matters in our teaching?

We realized that in order for this project to be successful we needed to worry less about helping the students move towards a completed product and instead focus on ensuring that they were able to spend each session adding to whatever artifact they were creating. The project needed to be more about doing and less about eventually having something that could be considered done. Within the school context this meant that we needed to truly serve more as a facilitator and less as a traditional teacher, focusing our attention on streamlining logistics and doing our best to just get out of students as they worked. Pedagogical flexibility, paying attention to how students are responding to the materiality of the space, holding an invitational stance as a facilitator of inquiry, and thoughtfully considering how this type of project might fit with other initiatives in a school are also important characteristics when introducing a university-led elective into the school schedule.

We agree with others who posit that a sense of agency would improve among all members of a school community if student achievement was reimagined and treated as a process of calibration, using more constructivist and inquiry-based pedagogies and research methods (Biesta, 2007; Chávez & Soep, 2005; Clark, 2011; Phelan, 2005; Sullivan, 2011). In our view, the aims of school accountability and constructivism do not need to be mutually exclusive.

Case 3: STEM Learning Through Robotics: Designing an After-School Program That Meets the Needs of Both Students and Researchers (Susan Lowes, Woonhee Sung, and Jung-Hyun Ahn)

In 2011, the Institute for Learning Technologies (ILT) at TC embarked on what it expected to be a one-time collaboration with the after-school program at the newly established TCCS to have TC doctoral students run a class that integrated technology into the curriculum. As of 2018, the program is still there, having evolved into a unique multi-year partnership that has allowed a changing group of TC doctoral students to develop a program of activities that engage students while at the same time researching their impact.

Although the specific activities and the technologies have changed over the 6 years, all of the research has been based in theories of embodied cognition, with a consistent focus on how to help students develop problem solving, programming, and computational thinking skills. Computational thinking—defined here as the ability to think like a computer scientist when confronted with a problem (Wing, 2006)—embraces such skills as problem identification, decomposition, sequential thinking, logical thinking, and pattern recognition. The ultimate goal is to develop curricular interventions that are not only based on theoretical understandings but that integrate technology in a way that is easy to implement across in-school and out-of-school settings.

The school's needs seemed simple: ILT would provide a program, offered one or two days a week, that had students using some form of technology in an engaging way. Although learning specific concepts was not a requirement, it was welcomed. In addition, it was understood that the activities would encourage students to think, to problem solve, and to collaborate.

ILT's needs seemed equally simple: TCCS would allow the ILT researchers to design the activities in such a way that their impact could be studied. The lead researchers have all been doctoral candidates funded through fellowships while their assistants have been a mix of doctoral and master's students funded through modest stipends provided by ILT and TC's Office of School and Community Partnerships.

The program has been a great success and ILT has been asked back year after year. From a research perspective, it has led to two doctoral dissertations, one completed (Sung, 2017) and one in progress, and to multiple presentations and journal publications (Ahn, Sung, & Black, 2018; Choi,

Marks, Lee, Ahn, & Black, 2015; Sung, Ahn, & Black, 2017; Sung, Ahn, Kai, & Black, 2017; Sung, Ahn, Kai, Choi, & Black, 2016). However, it should not be surprising that the process and the collaboration have not been so simple. This section looks at interaction between the after-school program and the research projects, seeking to address the question of how curriculum interventions can be developed and implemented in authentic settings while meeting both the researchers' needs and the needs of the school and its students.

Research: Theoretical approach and research questions. Embodied cognition researchers reject traditional views of cognition and argue that mental representations are constructed from bodily interactions with the physical world (Barsalou, 2008; Black, 2010; Wilson, 2002). The research suggests that embodied cognition can provide a powerful approach for helping novices learn abstract domains such as mathematics and physics by mapping physical actions and sensorimotor experiences to specific concepts in order to support learning (Bamberger & DiSessa, 2003; Barsalou, 2008; Segal, 2011). The first curriculum intervention, developed for the school's founding K–1 classes, focused on spatial problem solving and programming using the toy-like Beebots, with the students physically acting out (embodying) the path they wanted a Beebot to take. However, working with very young students proved difficult and the next iterations were with second, third, and fourth graders. The focus shifted to mathematical conceptual understanding (number line sense, numeracy, geometric knowledge) and the technology shifted to using Scratch and Hopscotch on an iPad. Embodied cognition was used to support the cognitive strategies needed to solve such problems as drawing a number line, estimating the distance between numbers, and drawing two dimensional geometric shapes. After the embodied activity, the students replicated the solution using Scratch or Hopscotch to make a virtual sprite solve the same problems. This was followed by the introduction of LEGO WeDo robotic construction sets to teach the concepts underlying simple machines. Embodied cognition was again used to map out the programming concepts. In other words, embodied cognition has been an integral part of all this work, whether the programming is done on an iPad or done in tandem with a physical robot.

The initial BeeBot research was one-on-one, with individual students pulled out of the class, but later the research was with the entire group. The research designs became more complex, with experimental and control groups and multiple conditions. For example, one study looked at two types of embodiment: direct embodiment, where the programmer (a student) acts out the coding himself or herself, and surrogate embodiment, where the programmer instructs the program (another student) to move in a certain way (Black, Segal, Vitale, & Fadjo, 2012). This introduced the concept of *computational perspective* (Patrick, Berland, & Martin, 2011) to the discussion of embodiment, since students had to take a programmer's point of view in order to provide verbal instructions to a surrogate. The group that took a computational perspective showed significantly higher gains in both mathematics and computational thinking skills (Sung, Ahn, & Black, 2017).

The research has recently shifted from having students program a robot to perform a series of actions to a more qualitative study of how students design and problem solve as they use the WeDo sets to construct swings, seesaws, and merry-go-rounds that can swing back and forth, move up and down, and rotate. These are open-ended activities with no one correct answer. In this study, there are two conditions, both of which take place at the planning stage: In one, students create flow charts showing the steps needed to create the final object, while in the other, students draw a picture of that object. In both our experimental and qualitative studies, including a planning stage has helped students begin the building process. In fact, preliminary results suggest that the students who construct the flow chart end up with more diverse final structures, possibly because they are focusing on the system—the way the parts fit together—while students who draw a picture tend to replicate the drawing.

Challenges and lessons learned. One of the benefits of working in an after-school setting is that it allows for activities that are not tied to specific learning mandates. The TCCS program required the curriculum to integrate technology in some way, involve some STEAM learning, and engage the students. Under these broad guidelines, it was acceptable if the activities changed depending on the researchers' interests, the technologies available, the classroom set-up, and the availability of such resources as laptops, robotic kits, and personnel. Nevertheless, over the course of the project a number of issues did arise. Most have been resolved due in large part to the good will and flexibility on the part of the school and the researchers. Some have necessitated changes in the design of the activities or the research. All have lessons for others about to embark on this type of collaboration.

It has been a logistical challenge to match students with specific iPads for research purposes. The students work in pairs, so changes in attendance mean that pairings change, and a student may use a different iPad each week. The solution has been to take screenshots of the work completed by each pair of students after each session as well as their system-generated activity records and then secure them separately on a password-protected hard drive.

There has also been an issue with the scheduling of the after-school program. TC classes are primarily in the afternoon and evening and are heavier on some days than others. This has made it a challenge to schedule the after-school program on a day that worked for the school—which has many other after-school activities—and is also at a time when there are enough TC students available to run the class. It has also been difficult to get a year-long commitment from the assistants since class schedules change from semester to semester. To resolve this, recruiting and training new assistants each semester has been integrated into the researcher's planning. The ability to pay a small stipend has also been an incentive.

Another issue has been that the school, responding to pressure from parents, decided after a few years that it wanted the program to focus on robotics, both because robotics was becoming popular and because the parents could see the results in the form of visible, tangible objects. This meant that the work on mathematics concepts, which was done entirely on an iPad, had to be replaced by a focus on building structures that integrated science concepts. LEGO WeDo sets were chosen because they were less expensive than the LEGO Mindstorms sets and the software would work on older iPads, but their functionality was more limited which in turn affected the design of the problems to be solved. However, this had the unintended consequence of forcing the researchers to develop activities that did not rely solely on technological tools—activities that have turned out to be usable in a wide range of other school circumstances.

One final logistical challenge has been the result of the success of the project. It became difficult to meet the needs of the school with the number of researchers and assistants available. Although the school would have liked more sessions, it was only possible to add one additional session each week due to the lack of capacity of a research-oriented program.

Other challenges have been related to the fact that the club is led by researchers, not teachers. For example, some of the doctoral student researchers had little direct classroom experience and struggled with controlling energetic younger students. This was resolved by bringing in research assistants with classroom experience, even if cognition and learning was not their research area. Even so, work with kindergarten and first grade students was dropped because it was too difficult to keep those students on task. This is an issue to be considered when working with younger students in a nonclinical setting.

At first, it was a challenge for the researchers to develop a program that lasted an entire year. Initially, the fall and spring terms were treated as separate curriculum interventions, to be repeated each semester, but the school wanted a year-long program. It took some rethinking and reconceptualizing, but the researchers found that they could scaffold the curriculum over the entire year if they first introduced basic programming concepts in Scratch and then explicitly built on them later by moving on to the WeDo construction kits.

Finally, some of the challenges were related to the fact that this was not only an after-school program but was also a research study. Perhaps the major challenge, from a research perspective, has been the inconsistency in attendance that is common in after-school settings. The TCCS after-school program is optional for students (who are often pulled out for family reasons), but enrollment is nevertheless capped. The result is that group membership can vary from week to week, leading to the possibility of a less than optimal number of students who have had the full experience and/or have taken all the assessments.

Neither the school, the students, nor ILT wanted after-school activities to replicate the school day. One challenge, therefore, was how to create assessments that would measure change, but that did not feel like tests to the students. The solution was the use of embedded performance assessments, with the first task used as a pre-test and the final task being a more difficult version of the previous tasks, integrating the same mathematical, science, or programming concepts.

It was also a challenge to create engaging activities that worked in both treatment and control conditions. One solution was to minimize the differences in what the students do. For example, in one of the early studies, where there was a no-embodiment, no-perspective taking control group, the students in the control group started with a paper-based activity, rather than an embodied one, but then moved on to the same iPad programming activity as the other groups. Since the paper-based activity took less time than the embodied activities, the control group finished the assessments earlier and they were then given a chance to experience the embodied activities. One issue here was the perception of the parents that their children are not having as much fun as their friends—for example, when their children came home and reported that they had to sit with a pencil and paper while their friends were moving around. Partly for this reason, the next studies did not have pure control groups, with the comparison instead being built on the use of different strategies.

Although programs that integrate hands-on problem solving with technology are attractive to students and to schools, not all schools are able to take advantage of the new technology whether this is because of limited funds, limited personnel, or lack of a usable curriculum. The curriculum developed by the ILT researchers was deliberately designed to prioritize pedagogy and reduce dependence on technological tools while still leading to gains in learning. The recent introduction of touch-gesture interfaces such as the iPhone and the iPad, along with simple programming languages like Scratch and LEGO WeDo, has made it easier to implement embodied learning environments (Sung, Ahn, & Black, 2017). The important point is that the program must be adaptable enough to take into account that any one—or all—of these constraints may change from one year to the next. If it is adaptable, then it seems likely that the program would work at other schools as well.

CONCLUSION

There are both opportunities and challenges in integrating computational thinking into the elementary school curriculum through a university-school partnership. Our experiences highlight some of the practical realities and also raise larger issues about roles that university partnerships might play in developing a culture of innovation in public schools. In reflecting on the lessons learned, we return to the three goals that guided our efforts: provide STEAM learning opportunities for students, model student-centered learning approaches for teachers, and study the relationship of computational thinking and learning. What we have learned relates to three themes raised in the recent NMC/CoSN Horizon Report (Freeman et al., 2017).

Advancing Progressive Learning Approaches Requires Cultural Transformation

Cultural transformation calls for a sustained, school-wide commitment coupled with ongoing support for teachers' efforts. In a university-school partnership, we believe that the impetus for a school-wide focus on innovation needs to come from the school rather than the higher education partners. However, in the ongoing policy debate over evaluating schools and teachers according to students' academic progress, it is no wonder that schools might be reluctant to adopt progressive learning approaches unless they can see the relationship of learning in this way to externally imposed measures of accountability. The *ISTE Standards for Students* (ISTE, 2018) may offer a useful framework for engaging in school wide technology integration with an "innovation" lens. Standards such as taking "an active role in choosing, achieving, and demonstrating competency" and "using digital tools to construct knowledge, produce creative artifacts, and make meaningful learning experiences for themselves and others" are relevant in all academic subjects (ISTE, 2018). It makes sense to start a transformation process by examining the similarities between *ISTE Standards for Students* (2018) and state learning standards.

In order to bring progressive learning approaches into the core of teaching and learning, schools and universities also need to invest in developing teachers. In our work, operational issues prevented us from creating voluntary pairings of classroom teachers with graduate student instructors although that approach might have encouraged greater mutuality, co-planning, and development of lesson designs that could move from the enrichment class to the teacher-led curriculum. We also realize that it takes multiple iterations for university and school-based co-instructors to learn from their experiences and consider ways to adapt the enrichment models to the regular classroom setting. Nonetheless, a distinct value of a university-school partnership lies in higher education's capacity to assist in the design and evaluation of learning as well as ongoing professional development to support teachers as they conceptualize and try out new approaches.

Learners Are Creators

In all three cases presented, project-based activities were highly motivating for students. Creativity expressed through digital channels is increasingly natural to this generation of students. As illustrated, most students eagerly embraced the agency required in open-ended learning opportunities. Yet, sometimes student agency in the regular classroom can be suppressed in order to accomplish goals dictated by current standardized testing and teacher evaluation systems (Ravitch, 2011; Varenne & McDermott, 1998). Unsurprisingly, teachers may favor the use of technologies that streamline acquisition of skills rather than those that foster open-ended learning. While it might be desirable for teachers to give students more freedom to create, schools can also make use of public-private partnerships to complement the core curriculum with opportunities for students to engage in creative problem solving, without the weight of accountability standards. If our experiences are representative, we can expect that advancing a culture of innovation through project-based STEAM learning is likely to remain peripheral to tested subjects for the time being.

Fluency in the Digital Realm Is More Than Just Understanding How to Use Technology

We discovered that the greater challenge of introducing technology-infused STEAM projects may lie in teaching students the skills needed to work together as a team. Our experiences are consistent with claims by Freeman et al. (2017) that computational thinking activities offer children opportunities to develop grit and character as they work through obstacles and frustrations. In our experiences, a student-driven project-based environment requires more of the teacher than technical and teaching skills; it also calls on them to mentor and coach students through personal doubts and interpersonal conflicts. These are familiar social-emotional learning goals in elementary education, but they have even more salience in the context of preparing students for the workforce. The *ISTE Standards for Students* (2018) may serve well in this domain, especially when aligned with schools' social-emotional curricula.

FINAL THOUGHTS AND RECOMMENDATIONS

Complementary, community-led STEAM programs are a good way to reach students directly even as schools work toward a blueprint for embedding technology in all aspects of learning. In the university-school partnership model, it is important for the university partner to set reasonable expectations for what to achieve with students and how to connect service to research as well as be prepared to address day-to-day implementation issues. Logistical glitches can be especially frustrating to college student instructors. Sharing hardware and software, occasional inaccessibility of teaching spaces, and changes to school schedules are common issues in schools, not unique to community-led programming. Yet, when enrichment classes are only held once a week, seemingly trivial logistical challenges have a greater impact on instructional time than they do on classes or equipment that can be shifted around from one day to the next. Orienting graduate students and tempering expectations can help mitigate some of the frustrations of being an outside partner.

Also, while providing curriculum that integrates technology in a thoughtful way is challenging in and of itself, the challenge is compounded by the demands of dissertation research. In our cases, the researchers often had to tailor their research plans and protocols to fit the environment. This is not necessarily a drawback of university-school collaboration but does require additional thought and a great deal of advance planning as well as the ability to adapt the research to the classroom rather than the other way around.

As noted by Freeman et al. (2017), across the country, students are "inventing, iterating, and collaborating regularly" and teachers are adapting to newer roles as facilitators of learning (p. 9). Universities need to consider how we can best prepare educators to guide students into our digital future. It is equally important to consider how we prepare building leaders and instructional coaches so they can lead schools toward a digital culture that promotes problem solving, critical thinking, creativity, and collaboration. University-school partnerships give preservice educators unparalleled opportunities to develop their knowledge, skills, and understandings about implementing technology-rich learning in authentic school settings. At the same time, our work with in-service teachers suggests that university partners must be willing to provide consistent support, exhibit patience, and be sensitive to teachers' and administrators' needs and priorities in order to build trust and value that is necessary to introduce more innovative pedagogical approaches.

Finally, if we are to optimize the potential of digital learning in schools, and move students beyond skill building toward broader aspects of computational thinking, we must also devise instruments that capture their learning. The university-school partnership model offers a context in which we can work together to examine the affordances of technology to develop student learning in ways that can be measured. Universities can advance the momentum toward progressive and innovative learning environments by building the evidence base that demonstrates how such approaches help schools meet their student achievement and college/career readiness goals.

In summary, there have been tremendous advances in teaching computational thinking in K-12 education. If we want to capture the greatest potential of technology for teaching problem solving, critical thinking, creativity, and collaboration in elementary education, then scholars, educational policy makers, and schools should continue to explore the alignment of standards, goals, and assessment practices related to digital learning.

REFERENCES

Ahn, H., Sung, W., & Black, J. B. (2018, April). *Using embodied instruction and programming language formats in young children's debugging activities.* Paper presented at the annual conference of the American Educational Research Association, Washington, DC.

Bamberger, J., & DiSessa, A. (2003). Music as embodied mathematics: A study of a mutually informing affinity. *International Journal of Computers for Mathematical Learning, 8,* 123–160.

Barsalou, L. W. (2008). Grounded cognition. *Annual Review of Psychology, 59,* 617–645. doi.org/10.1146/annurev.psych.59.103006.093639

Biesta, G. (2007). Why "what works" won't work: Evidence-based practice and the democratic deficit in educational research. *Educational Theory, 57*(1), 1–22. doi.org/10.1111/j.1741-5446.2006.00241.x

Black, J. B. (2010). An embodied/grounded cognition perspective on educational technology. In M. S. Khine & I. Saleh (Eds.), *New science of learning: Cognition, computers and collaboration in education.* New York, NY: Springer.

Black, J. B., Segal, A., Vitale, J., & Fadjo, C. L. (2012). Embodied cognition and learning environment design. In D. Jonassen & S. Land (Eds.), *Theoretical foundations of learning environments.* New York, NY: Routledge.

Chávez, V., & Soep, E. (2005). Youth radio and the pedagogy of collegiality. *Harvard Educational Review, 75*(4), 409–434. doi.org/10.17763/haer.75.4.827u365446030386

Choi, A., Marks, J., Lee, A., Ahn, J., & Black, J. B. (2015, April). *The effect of cognitive embodiment on children's problem solving in robotics education.* Paper presented at annual meeting of the American Educational Research Association, Chicago, Illinois.

Clark, A. (2011). Multimodal map making with young children: Exploring ethnographic and participatory methods. *Qualitative Research, 11*(3), 311–330. doi. org/10.1177/1468794111400532

Freeman, A., Adams Becker, S., Cummins, M., Davis, A., & Hall Giesinger, C. (2017). *NMC/CoSN horizon report: 2017 K–12 edition.* Austin, TX: The New Media Consortium.

Freire, P. (1970). *Pedagogy of the oppressed* (Myra Bergman Ramos, Trans.) New York, NY: Continuum.

International Society for Technology in Education. (2018). *ISTE standards for students.* Retrieved from https://www.iste.org/standards/for-students

ISTE (International Society for Technology in Education). (2011). *Operational definition of computational thinking for K–12 education.* Retrieved from http://www.iste.org/docs/ct-documents/computational-thinking-operational-definition-flyer.pdf

Kanellopoulos, P. A. (2012). Envisioning autonomy through improvising and composing: Castoriadis visiting creative music education practice. *Educational Philosophy and Theory, 44*(2), 151–182. doi.org/10.1111/j.1469-5812.2010.00638.x

Lewis, C., & Moje, E. B. (2003). Sociocultural perspectives meet critical theories: Producing knowledge through multiple frameworks. *The International Journal of Learning, 10,* 1980–1995.

Maeda, J. (2013, March). *Live interview with John Maeda.* Retrieved from http://stemtosteam.org/events/congressional-steam-caucus/

National Science Foundation. (n.d.). *CS for all.* Retrieved from https://www.nsf.gov/news/special_reports/csed/csforall.jsp

Patrick, C., Berland, M., & Martin, T. (2011). *Allocentrism and computational thinking.* Paper presented at the Ninth International Conference on Computer-Supported Collaborative Learning, Hong Kong, China.

Phelan, A. M. (2005). A fall from (someone else's) certainty: Recovering practical wisdom in teacher education. *Canadian Journal of Education/Revue Canadienne de l'Education,* 339–358. doi.org/10.2307/4126474

Ravitch, D. (2011). *The death and life of American school review: How testing and choice are undermining education.* New York, NY: Basic Books.

Segal, A. (2011). *Do gestural interfaces promote thinking? Embodied interaction: Congruent gestures and direct touch promote performance in math.* New York, NY: Columbia University.

Sousa, D. A., & Pilecki, T. (2013). *From STEM to STEAM: Using brain-compatible strategies to integrate the arts.* Thousand Oaks, CA: Corwin Press.

Streim, N. (2016). Building a university-assisted school through community collaboration. In J. Slater, R. Ravid, & M. Reardon (Eds.), *Building and maintaining collaborative communities: Schools, university and community organizations.* Charlotte, NC: Information Age Publishing.

Sullivan, G. (2011). The culture of community and a failure of creativity. *Teachers College Record, 113*(6), 1175–1195.

Sung, W. (2017). *The impact of embodiment and computational perspective-taking practice on young children's mathematics and programming ability* (Doctoral dissertation). Retrieved from ProQuest (10286683).

Sung, W., Ahn, J., & Black, J. B. (2017). Introducing computational thinking to young learners: Practicing computational perspectives through embodiment in mathematics education. *Technology, Knowledge and Learning, 22*(3), 443–463. doi.org/10.1007/s10758-017-9328-x

Sung, W., Ahn, J., Kai, S., & Black, J. B. (2017). Effective planning strategy in robotics education: An embodied approach. In P. Resta & S. Smith (Eds.), *Proceedings of Society or Information Technology & Teacher Education International Conference 2017* (pp. 754–760). Chesapeake, VA: Association for the Advancement of Computing in Education (AACE).

Sung, W., Ahn, J., Kai, S. M., Choi, A., & Black, J. B. (2016). Incorporating touch-based tablets into classroom activities: Fostering children's computational thinking through iPad-integrated instruction. In D. Mentor & IGI Global (Eds.), *Handbook of research on mobile learning in contemporary classrooms* (pp. 378–406). Hershey, PA: IGI Global.

Teachers College Community School. (n.d.). Retrieved from https://www.tccsps517.org/about/

Varenne, H., & McDermott, R. (1998). *Successful failure.* Boulder, CO: Westview Press.

Verdiner, M. (2018, May 22). Personal interview with N. Streim.

Wilson, M. (2002). Six views of embodied cognition. *Psychonomic Bulletin & Review, 9*(4), 625–636. doi.org/10.3758/BF03196322

Wing, J. M. (2006). Computational thinking. *Communications of the ACM, 49*(3), 33–36. doi.org/10.1145/1118178.1118215

PART III

BUILDING ACADEMIC PROFICIENCY

CHAPTER 6

USING TECHNOLOGY TO FACILITATE P–20 PARTNERSHIPS IN RURAL COMMUNITIES

Elizabeth E. Smith
The University of Tulsa

Heather Young and Vinson Carter
University of Arkansas

Broadly speaking, partnerships between P–12 schools and institutions of higher education fit into a larger group of partnerships referred to as P–20 partnerships. Technology provides an opportunity to reconceptualize P–20 partnerships outside the traditional uses of web-based partnerships or video-conferencing for concurrent courses. This chapter focuses on one partnership between a university and a rural elementary school that used both new and traditional types of technology to promote STEM content understanding and career exploration.

Integrating Digital Technology in Education:
School–University–Community Collaboration, pp. 137–149

Historically, P–12 schools and institutions of higher education have operated largely independently of one another (Venezia & Kirst, 2005). The gap between P–12 schools and postsecondary institutions consistently grew throughout the 20th century and is recognized as a factor in a number of troubling issues, including low high school graduation and college-going rates for minority students, high percentages of students who require remedial courses in college, and declining college readiness overall (Carnevale, Smith, & Strohl, 2013; Kena et al., 2014). An increasingly popular approach to address this gap is the creation of programs that encourage cooperation, collaboration, and coordination between the two systems (Cox-Peterson, 2011; Gianneschi & Fulton, 2014). Typically, P–20 partnerships include two or more parties who agree to work together to improve student learning (Cox-Peterson, 2011). P–20 partnerships often exist in a "third space" (Martin, Snow, & Franklin Torrez, 2011, p. 299; see also Fisher & Many, 2014) in which those who engage in such partnerships act to span boundaries. Wilbur and Lambert (1995) reviewed 2,322 partnerships and found that they typically take four forms: (a) programs and services for students; (b) programs and services for educators; (c) articulation, development, and evaluation of curriculum and instruction; and (d) restructuring.

Research suggests that experiences in mathematics and science as early as pre-K and into early elementary school may affect student learning and attitudes about STEM (National Association for the Education of Young Children & National Council of Teachers of Mathematics, 2010; National Science Board, 2010). Despite the lucrative potential of employment in STEM and the seemingly boundless opportunities in these fields, many young people remain reluctant to enter career fields that require a background in STEM. Murphy (2011) suggested that children are natural problem-solvers at birth. They try to make sense of the world around them through exploration: touching, tasting, building, dismantling, and creating. Murphy noted that children do not consider this education, but truly enjoyable learning instead. However, by the time students reach the fourth grade, approximately 30% have lost interest in STEM, and by eighth grade, another 20% have decided that STEM is not related to their future plans (Murphy, 2011). Oakes, Leone, and Gunn (2006) noted that even those students who do not turn their backs on STEM often enter postsecondary programs without a clear understanding of the field, its practice, or its effects on society. Given this startling information, it is imperative that P–12 teachers help students understand STEM's significance. This understanding was explored further in this chapter which focuses on a collaborative P–20 partnership involving university faculty, university students, elementary school faculty, and elementary students.

ADOPT-A-CLASSROOM

The Education Renewal Zone at the University of Arkansas in Fayette-ville, Arkansas created Adopt-A-Classroom (AAC) to facilitate connections between higher education faculty and local P–12 teachers (Smith, Kindall, Carter, & Beachner, 2016). AAC was created in response to P–12 adminis-trators who asked university administrators for support as teachers sought to deepen their content knowledge. University faculty agreed to visit their P–12 partner classrooms at least once a month to teach or co-teach and connect teachers and P–12 students to the resources available at the uni-versity. While the program was designed specifically to benefit university faculty and P–12 teachers, P–12 students also took part in and benefitted from the partnerships. Partner schools that participated in all AAC part-nerships were a mix of rural and suburban schools and a majority of the students involved were from low-income backgrounds.

PROGRAM OVERVIEW

Rural schools often lack the financial resources and infrastructure to offer students many of the tools regarded as 21st century building blocks (Sundeen & Sundeen, 2013). However, one study found that despite these limited resources, rural teachers held more positive attitudes about the integration of technology than did their non-rural counterparts (Howley, Wood, & Hough, 2011). This same study posited that because of these teachers' positive attitudes, students in rural locations could benefit greatly from integrating technology as an effective learning tool.

An integrated STEM design challenge was created for this AAC part-nership and implemented in a Grade 4 classroom in a rural school. The design challenge required students to create a play based on an historical event and used both new and traditional technologies to tell their stories. Students were provided with a variety of technologies and asked to create marionettes and a stage on which the marionettes performed the play they created. Students developed their plays using several themes, including incorporating informational and fictional texts to explore historical events, applying scientific inquiry approaches and engineering designs, learning to use puppetry and anthropometrics to tell the stories, becoming familiar with simple tools and processes, and honing presentation skills.

School Partner

The Grade 4 classroom involved in this partnership was located in a rural intermediate school which housed students in Grades 4 and 5. The school had an enrollment of 369 students during the partnership with an average class size of 24. The school included 44% free and reduced lunch participants, 5% English language learners, and 13% of students who qualified for special education. The race/ethnicity of the partnership school was reported to be 1% Native American, 3% Black/African American, 3% two or more races, 8% Hispanic/Latino, and 85% Caucasian. The participating teacher who volunteered for the AAC project was paired with two university faculty members. This was the second year in which the classroom teacher participated in the AAC, but the first year in which this specific team collaborated. The classroom teacher was National Board Certified, served as the school's STEM facilitator, and had been teaching Grade 3 or Grade 4 for 14 years. She taught primarily science and social studies, however the group of students involved in this collaboration also included her homeroom students.

Partnership Design

The following essential questions were posed to the students at the beginning of the project: (1) How does the past affect the present? (2) How can a script for a play be developed from a children's book? (3) How can marionettes and a set be designed to animate a written story that will engage the audience and narrate the story? The students in each class were asked to work as members of an engineering design team to bring an historical informational text to life through the use of puppetry, set design, and development of a puppet drama or a marionette play. Students were given the opportunity to use traditional technologies, such as table-top vises and hand saws, as well as new technologies through the Puppet Pals© iPad application (app). Both types of technologies were integral to their learning and necessary to complete the project.

At the beginning of the project, student groups (of three or four) identified an historical event from a children's book to use as the storyline for the marionette play. The books were based on the current topic they were studying in social studies: the American Revolution. Students then worked in teams to write a script for the marionette play. Next, the student groups used the Puppet Pals© iPad app to design a prototype play to develop their characters and refine their draft script. The students then brought their prototypes from the iPad to life by developing marionette characters using simple materials including newspaper, pipe-cleaners, masking tape, popsicle sticks, fishing line, tempera paint, and other recycled materials. The students also used recycled materials (including cardboard boxes and wood) to design a set and props. Each student was required to design a marionette and participate as a character with a speaking part.

Students learned about the engineering design process as the foundation for the project and began to understand the way engineers and scientists use the design process to solve problems and develop solutions. They also were taught that brainstorming and documentation are important elements of the design process. To facilitate brainstorming and documentation, students used an engineering design journal throughout the process of creating their plays to conduct research, gather information, document ideas, and make plans. Students defined the problems that they were asked to solve and then identified the products that their team would need to develop to solve the design challenge. In addition, they identified the historical event and text to be used as the storyline for the marionette play. Students also developed sketches for each scene included in the marionette play.

After choosing an historical text and writing the script for their play, students created a storyboard using examples of comic books and graphic novels. During the storyboarding process, students determined the setting for their plays and considered a variety of questions, including what the background would look like and what their marionettes might wear. Next, they used the Puppet Pals© iPad app to create a prototype video of their play. Puppet Pals© is a digital learning tool that can be used to create a video and audio recording on an iPad. Within the application, users can add music, backgrounds, characters, voice recordings, and save and share a completed performance video. Upon completion of their virtual sets, students designed physical sets for the play using recycled and repurposed materials. Students were asked to design marionettes that reflected the characters in their selected book and were reminded about the importance of measuring to achieve the appropriate scale and dimensions for their set design. Students also learned about the proper use of hand tools, such as a miter saw and drill press and used those tools to create their marionettes and sets.

Role of University Faculty, P–12 Teachers, and Preservice Teachers

The two members of the university faculty who participated in this partnership (one a professor in childhood education and the other a professor in STEM education) incorporated the project into two classes they were teaching (Children's Literature and Creativity and Innovation in STEM Education) so that university preservice teachers also could be involved in the partnership. Prior to beginning the classroom sessions, the partners met to orient them to the students in the classroom and discuss the state and national standards that the project would address. They also discussed the technology available in the participating school and the students' familiarity with the engineering design process. The university faculty and pre-service teachers visited the classroom five times over an academic year and during these visits, the two faculty members acted as co-teachers, while the preservice teachers supported the small groups of Grade 4 students. Once the projects were completed, the students visited the University of Arkansas and presented their marionette plays to an audience including other P–12 students, parents, professors, and college students in a children's literature course.

OUTCOMES

The university faculty designed a pre- and posttest influenced by the Friday Institute for Educational Innovation (FI, 2012) to investigate the elementary student participants' attitudes regarding STEM, the engineering design process, and potential careers in STEM disciplines. The assessment was developed based on a literature review and was consistent with the student learning objectives for our partnership which was designed to give elementary students hands-on experience of a STEM project, improve their understanding of science, technology, mathematics, and engineering, and help them to understand better what scientists and engineers do as part of their profession.

The pre- and posttest we developed consisted of two scales with a total of 15 questions—one with seven items focused on students' perceptions of their STEM abilities and another with eight items focused on students' plans for the future and their perceptions about STEM careers. In addition, the students were invited to draw a picture representing what scientists and engineers could make that would make a difference in their lives. The 15 questions that comprised the two scales were three-option Likert-scaled questions that used emoticons to represent disagree (1), unsure (2), and agree (3). Sixteen fourth-grade students at the rural partner school completed the pre- and posttest, six males and 10 females.

Following completion of the posttest, means of each of the 15 questions were examined to investigate differences. Table 6.1 lists the pre- and posttest means and differences for each of the 15 questions. Differences from pre- to posttest ranged from –0.0625 to 0.625. On the posttest, the mean responses to two questions decreased slightly: "I am creative" (–0.08) and "I like to start problems I don't know the answer to" (–0.06). The largest increase from pre- to posttest was in "I am good at engineering," with an increase of 0.63. The three other questions that showed the largest increases from pre- to posttest were "I am good at science," "I am good at math," and "I would like to be an engineer." There was an overall increase in student responses from the pretest ($M = 2.53$) to the posttest ($M = 2.72$).

Table 6.1.
Pre- and Posttest Differences in Means

Question	Pretest Mean	Posttest Mean	Difference
I am good at science.	2.63	2.94	0.31
I am good at math.	2.56	2.88	0.32
I am good at engineering.	2.06	2.69	0.63
I like learning how things work.	2.69	2.87	0.18
I am creative.	2.89	2.81	-0.08
I like to start problems I don't know the answer to.	2.13	2.07	-0.06
I like building things.	2.75	3.00	0.25
I would like to be a scientist.	2.27	2.56	0.29
I would like to be an engineer.	2.06	2.38	0.32
I would like a job where I invent things.	2.44	2.63	0.19
I would like to design machines that help people.	2.75	2.75	0.00
Scientists help make people's lives better.	2.80	2.94	0.14
Engineers help make people's lives better.	2.80	2.81	0.01
I know what scientists do for their jobs.	2.56	2.75	0.19
I know what engineers do for their jobs.	2.50	2.75	0.25

A paired t test was performed to examine the significance of the difference in means between the pretest and posttest. The difference between the means from the pretest to the posttest ($M = -0.196$, $SD = 0.170$) was significant at the 95% confidence interval ($p < .001$).

Table 6.2.
Results of Paired *T* Test for Difference Between Means

M	SD	*t* test
−0.196	0.170	−4.732**

** $p < .001$, two-tailed

While not limited to the faculty involved in this specific partnership, an evaluation of the entire AAC program found that both university faculty and P–12 teacher reported positive effects of participation on their teaching (Smith et al., 2016). University faculty reported that partnering with the P–12 teachers provided them with examples of innovative teaching and made them better instructors. P–12 teachers reported that their participation spurred them to rethink their teaching and provided them with new ideas for classroom activities and projects.

DISCUSSION

The results from this 1-year pilot partnership indicate that introducing Grade 4 students to STEM projects can have a positive effect on their perception and understanding of STEM content areas and their early plans to pursue careers in STEM fields. Following completion of the project, the rural students who participated reported higher levels of self-efficacy in science, mathematics, and engineering. Further, their stated interest in pursuing careers in science and/or engineering increased. While this was only a 1-year pilot partnership, it provided a template for other potential research-practice partnerships (RPPs) designed to improve attitudes about STEM in rural elementary students. RPPs are long-term partnerships to improve schools that connect university researchers with P–12 schools (Coburn & Penuel, 2016; Coburn, Penuel, & Geil, 2013; Luter & Kronick, 2017). Accordingly, a school district could consider implementing the project-based model described in this chapter at a low-cost and potentially have a strong influence in improving elementary students' perceptions of STEM content areas and careers. The cost for physical materials was minimal; the largest expense was in human capital for the time the university faculty and P–12 teachers spent designing and implementing the project.

Translating this pilot partnership to a long-term RPP model would benefit all partners. The effects of the pilot partnership could be compounded by creating similar projects for all elementary grades that include the use of both traditional and new technologies in partnerships with university faculty. The general structure of the program (pairing university faculty with P–12 teachers and using STEM design principles to bring historical events to life) can be used to create projects at any grade level and provide P–12 students with the opportunity to use a variety of technologies. A district-wide, longitudinal approach would allow students to build on their content knowledge and use of STEM technologies over time, perhaps moving from simple to more complex tools. For example, during the project described in this chapter, students learned to use age-appropriate hand tools such as mini hacksaws with miter boxes and hand drills. The students also learned how to use different features on the iPad. Creating a long-term project in an RPP approach would provide students with a foundation to use increasingly more complex tools as they progress through school. Additionally, using a multiyear RPP approach would give students the opportunity to augment their STEM knowledge and enhance their confidence in using STEM technologies each year.

Another advantage of translating the AAC partnership described in this chapter to an RPP is the ability to track students' attitudes about STEM content longitudinally. If an RPP model was employed, university faculty could follow-up with P–12 teachers and students in subsequent years to investigate whether the partnership had long-term effects. Long-term partnerships in which the same teachers are included also have the potential to increase STEM expertise among participating P–12 teachers. By involving four distinct groups of stakeholders (P–12 teachers, their students, university faculty, and university students), AAC partnerships can lead to establishing long-term collaboration that have the potential to impact the entire P–20 system in a district.

Beyond the student perceptions assessed from this pilot study, further studies need to investigate the way this intervention affects students' STEM career trajectories. As mentioned above, a longitudinal approach would allow researchers to investigate potential long-term impacts of this partnership. Future studies should also examine results from the pre- and posttests in light of student demographic data to see if there are differences between the program's impact on students from difference backgrounds. Future research could also examine the experience of pre-service teachers involved in similar partnerships. Does their participation influence their perception of teaching STEM?

CHALLENGES AND OPPORTUNITIES

One challenge when using technology in rural P–20 partnerships is arriving at a shared definition of technology. There is a common misconception in education that the definition of technology entails solely the use of computers, or higher levels of technology, such as computational and communication devices (Daugherty, 2010; National Research Council, 2012;

Sanders, 2009). The Standards for Technological Literacy (ITEEA, 2007 [No reference]) define technology in the context of

> how people modify the natural world to suit their own purposes. From the Greek word techne, meaning art or artifice or craft, technology literally means the act of making or crafting, but more generally refers to the diverse collection of process and knowledge that people use to extend human abilities and to satisfy human needs and wants. (p. 2)

Further, the National Research Council (2012) has proposed that "technologies result when engineers apply their understanding of the natural world and of human behavior to design ways to satisfy human needs and wants" (pp. 11–12). When considering technological literacy in P–12 classrooms, then, technology does not just include the use of computers, iPads, or interactive white boards. Instead, teachers and students can use a variety of materials to create or craft, employing both old and new technologies to encourage students to mold their environment and solve real-world problems. The AAC partnership enabled students to create and craft while making sense of informational texts. Students and teachers embodied the definitions of technological literacy by using tools to extend their understanding of a topic. Creating prototypes and building sets and marionettes alongside writing a script based on texts gave students an opportunity to more deeply examine the literary concepts of setting, character, and plot development.

While P–12 partners for this project were familiar with iPads and other types of computer-based technology, they were unaccustomed to using traditional types of technology including basic hand tools. The P–12 teacher and students were very willing to learn how to use these tools after safety concerns were addressed. The partnership described in this chapter broadened the definition of technology in the P–12 school by introducing the students and teachers to traditional technologies that typically are unavailable to them. In this case, the challenge of arriving at a shared definition of technology was an opportunity for P–12 partners to encounter new types of technology and for the university partners to observe the way elementary students can use traditional types of technology.

The availability of resources was a challenge in completing this project and frequently is a challenge when developing P–20 partnerships in rural areas. While the project initially was designed for a classroom with a set of iPads, the P–12 teacher had only one, so university faculty brought enough additional iPads for each group to have one. This created challenges within the groups as students had to take turns using the iPads or have other responsibilities that did not include using them. The partner school also did not have the traditional technology available, so the university faculty brought saws and drills (and appropriate safety equipment) from the university's STEM laboratory.

Partnerships with rural schools provide university faculty and students with the opportunity to understand the challenges rural students and teachers face. The university faculty involved in this partnership had initial assumptions about the technology available at the rural school based on their previous experience with suburban schools, but discovered quickly that their rural partners had less access to technology than they expected. As Smith et al. (2016) described, university faculty in AAC partnerships reported gaining an understanding of the challenges faced by P–12 teachers and a better understanding of the development students experience prior to college.

Regardless of whether the use of technology is included, P–20 partnerships in rural areas provide a link between P–12 schools and higher education that can benefit P–12 students and teachers and university students and faculty as well. Students in rural schools may consider college to be beyond their reach but having contact with university faculty may encourage them to see college as an option. Partnerships between rural schools and universities not only benefit the P–12 participants. The university students who participated in this partnership were pre-service educators at the time and they benefitted from engaging with students in a rural classroom by supporting the small groups of P–12 students as they designed their plays and built their sets. Importantly, AAC was framed as a partnership seeking simultaneous renewal, acknowledging that the university faculty and university students had much to learn from the P–12 partners (Goodlad, 1994).

CONCLUSION

P–12 students who participated in this pilot partnership reported an increased efficacy in science, mathematics, and engineering at the conclusion of the one-year project. In addition, P–12 students indicated an increased interest in pursuing STEM careers. Partnership programs similar to AAC also provide P–12 teachers and university faculty opportunities for collaboration, content integration, and professional learning. Thus, P–20 partnerships in STEM have the potential to impact all aspects of the P–20 system positively.

REFERENCES

Carnevale, A., Smith, N., & Strohl, J. (2013). The road to recovery: Projecting U.S. job growth and higher education demand through 2020. *Community College Journal, 84*(3), 26–29.

Coburn, C. E., Penuel, W. R., & Geil, K. (2013). *Research-practice partnerships at the district level: A new strategy for leveraging research for educational improvement.* New York, NY: William T. Grant Foundation.

Coburn, C. E., & Penuel, W. R. (2016). Research–practice partnerships in Education: Outcomes, dynamics, and open questions. *Educational Researcher, 45*(1), 48–54. https://doi.org/10.3102%2F0013189X16631750

Cox-Peterson, A. (2011). *Education partnerships: Connecting schools, families, and the community.* Los Angeles, CA: SAGE.

Daugherty, M. K. (2010). The 'T' and 'E' in STEM. In ITEEA (Ed.), *The overlooked STEM imperatives: Technology and engineering* (pp. 18–25). Reston, VA: ITEEA.

Fisher, T. R., & Many, J. E. (2014). From PDS classroom teachers to urban teacher educators: Learning from professional development school boundary spanners. *School-University Partnerships, 7*(1), 49–63.

Friday Institute for Educational Innovation. (2012). *Student attitudes toward STEM: The development of upper elementary school and middle/high school student surveys.* Raleigh, NC: Author. Retrieved from https://eval.fi.ncsu.edu/wp-content/uploads/2013/03/Student-Attitudes-toward-STEM-ASEE13-ERM-Final.pdf

Gianneschi, M., & Fulton, M. (2014). *A cure for remedial reporting chaos: Why the U.S. needs a standard method for measuring preparedness for the first year of college.* Denver, CO: Education Commission of the States. Retrieved from http://www.ecs.org/docs/Cure-for-Remedial-Reporting-Chaos.pdf

Goodlad, J. (1994). *Educational renewal: Better teachers, better schools.* San Francisco: CA, Jossey-Bass.

Howley, A., Wood, L., & Hough, B. (2011). Rural elementary school teachers' technology integration. *Journal of Research in Rural Education, 26*(9), 1–13.

International Technology Education Association. (2007). *Standards for technological literacy: Content for the study of technology* (3rd ed.). Reston, VA: Author.

Kena, G., Aud, S., Johnson, F., Wang, X., Zhang, J., Rathbun, A., . . . Kristapovich, P. (2014). *The condition of education 2014.* Washington, DC: U.S. Department of Education, National Center for Education Statistics. Retrieved from https://nces.ed.gov/pubs2014/2014083.pdf

Luter, G., & Kronick, R. F. (2017). Community/school/university partnerships as catalyzing reform in districts and across the state: The university assisted community schools project in Knoxville, TN. In R. M. Reardon & J. Leonard (Eds.), *Exploring the community impact of research-practice partnerships in education* (pp. 117–146). Charlotte, NC: Information Age.

Martin, S. D., Snow, J. L., & Franklin Torrez, C. A. (2011). Navigating the terrain of third space: Tensions with/in relationships in school-university partnerships. *Journal of Teacher Education, 62*(3), 299–311. doi: 10.1177/0022487110396096

Murphy, T. (2011, August 29). STEM education—It's elementary. *US News and World Report.* Retrieved from http://www.usnews.com/news/articles/2011/08/29/stem-education--its-elementary

National Association for the Education of Young Children & National Council of Teachers of Mathematics. (2010). *Early childhood mathematics: Promoting good beginnings.* Retrieved from https://www.naeyc.org/files/naeyc/file/positions/psmath.pdf

National Research Council, Committee on a Conceptual Framework for New K–12 Science Education Standards. (2012). *A framework for K–12 science education: Practices, crosscutting concepts, and core ideas.* Washington, DC: The National Academies Press.

National Science Board. (2010). *Preparing the next generation of STEM innovators: Identifying and developing our nation's human capital.* Retrieved from https://www.nsf.gov/nsb/publications/2010/nsb1033.pdf

Oakes, W. C., Leone, L. L., & Gunn, C. J. (2006). *Engineering your future: An introduction to engineering.* St. Louis, MO: Great Lakes Press.

Sanders, M. (2009). Integrative STEM education: Primer. *The Technology Teacher, 68*(4), 20–26.

Smith, E., Kindall, H. D., Carter, V., & Beachner, M. (2016). Impact of Adopt-A-Classroom partnerships between K–12 and university faculty. *School Community Journal, 26*(1), 163–181.

Sundeen, T. H., & Sundeen, D. M. (2013). Instructional technology for rural schools: Access and acquisition. *Rural Special Education Quarterly, 32*(2), 8–14. https://doi.org/10.1177%2F875687051303200203

Venezia, A., & Kirst, M. W. (2005). Inequitable opportunities: How current education systems and policies undermine the chances for student persistence and success in college. *Educational Policy, 19*(2), 283–307. doi:10.1177/0895904804274054

Wilbur, F., & Lambert, L. (Eds.). (1995). *Linking America's schools and colleges* (2nd ed.). Washington, DC: American Association for Higher Education.

CHAPTER 7

TECH INEQUITY

Preservice Teachers Combating the Digital Divide in an Urban School- and Community-Based Immersion Program

Abiola Farinde-Wu
University of Massachusetts, Boston

Aaron J. Griffen
DSST Public Schools, Denver, Colorado

In many urban schools and contexts, technology availability, access, and use are limited. Seeking to combat this digital issue, researchers at one, large, urban institution in the northeast region of the United States conceptualized and implemented the Students Success (SS) tutoring and mentoring program, an initiative geared toward assisting underserved students of color. Highlighting SS's technology component and with themes from the NMC/CoSN Horizon Report in mind, this case study examines how prospective preservice teachers in an urban school- and community-based immersion program served as resources by promoting equity in the integration of digital technology in teaching and learning. Although in-school technology use occurred at their urban school site, technology was outdated, often restricted in terms of access, and unresponsive to high school students' learning needs.

Integrating Digital Technology in Education:
School–University–Community Collaboration, pp. 151–171

In this school context and with aid from their prospective preservice teachers, high school students utilized university resources to construct technology projects focused on strengthening their math and English comprehension. To this end, the process of constructing students' technology projects captured how a school-university-community collaboration positively influenced the digital divide and the technology experiences of underserved students of color.

In today's technologically driven society, computers and the Internet are ubiquitous. The ubiquity of such knowledge technologies supports the illusion that all members of society have equal access to and ownership of these affordances. In actuality, there exists a technology disparity between various racial and ethnic groups (Jackson et al., 2008), which is often dictated by socioeconomic status. This social and economic inequality is referred to as the digital divide, the gap between those who have access to information and communication technology and those who do not (Organization for Economic Co-operation and Development [OECD], 2002). For instance, when examining technology use among White and Black adults, Gorski (2008) shared that "despite the popular belief that identity-based discrepancies in physical access to computers and the Internet are disappearing, substantial gaps remain" (p. 352). In his study, Gorski noted that 70% of White adults in the United States use the Internet, whereas only 57% of African Americans are online. Also, as mentioned above, this gap is influenced by socioeconomic status, where homes with median or high household incomes have greater access to the Internet than low-income households.

Although there is evidence of a diminishing digital divide (OECD, 2004), in many urban schools and contexts technology availability—the opportunity to access and use technology—is limited (Kormos, 2018; Warschauer, Knobel, & Stone, 2004). Furthermore, Dolan (2016) outlined additional issues confronting K–12 students when they attempt to use technology. She explained that these issues frame students' experiences with technology, redefining the current digital divide definition and expanding it to include digital inequity. She highlighted

Internet connectivity and bandwidth capabilities; availability of software; students' and teachers' knowledge and skills to use the available technology; the influence of mobile technology; and the impacts of limiting factors such as poverty, lack of teacher training, and cultural misunderstandings between students and teachers. (p. 16)

Considering digital inequities among racial/ethnic student groups from different school contexts, researchers at one, large, urban institution in the northeast region of the United States conceptualized and implemented the Student Success (SS) tutoring and mentoring program—an initiative geared toward assisting underserved students of color in an urban school. Based on the twin assumptions that the "widespread use of technology does not translate into equal learner achievement" for all students, and that "there is no replacement for good teaching—the role is just evolving" (Freeman, Adams Becker, Cummins, Davis, & Hall Giesinger, 2017, p. 4), the purpose of this case study is to examine the technology component of the Student Success program. As Freeman, Adams Becker, Cummins, Davis, and Hall Giesinger (2017) explained, while technology assists users, it does not alone negate issues of socioeconomic status, race, ethnicity, and gender that influence student engagement and performance. Aware that technology facilitates but is not the sole determinant of positive learning experiences and achievement for students, the following research question guided this study: How do prospective preservice teachers in an urban school- and community-based immersion program serve as resources to promote equity in the integration of digital technology in teaching and learning?

LITERATURE REVIEW

There is a digital divide that exists among all other divides and gaps. Prioritizing one gap or divide over another, such as the achievement gap versus the opportunity gap or the digital divide versus the equity gap, leads to one of many conclusions. That is, there is no single divide or gap in education but a series of complex factors contributing to and exacerbating educational inequalities (Warschauer et al., 2004). The digital divide is only one of many proposed inequities that has been compared to and suggested as resulting from all the other gaps and divides. Although there is a growing population of digital natives,[1] levels of competence in computer literacy and skills remain low (Lei, 2009). Regardless of the push for schools to become technology based through 1:1 initiatives, where all students are provided a laptop and blended learning models, there is a population of learners and teachers who struggle to utilize technology to its fullest capacity, leading to further inequities in school technology use.

Systemic Inequity

The issue that continues to emerge regarding the digital divide is equity. Warschauer et al. (2004) asserted that "low-SES schools [employ] nearly five times as many teachers without full credentials" (p. 568). This disparity translates into disparity in teachers' technology skills, implementation, efficacy, and beliefs, widening the digital divide (Wilson, 2014). Despite improvements in training and allocation of resources to facilitate technology implementation in urban schools, a lack of opportunity to be fully successful with technology in a timely manner emerges for students of color and students from low socioeconomic backgrounds, and the teachers who work with these populations (Hess & Leal, 2001; Kormos, 2018; Mouza, 2008; Staples, Pugach, & Himes, 2005). According to Neuman (1991), "not only minority, disadvantaged, and inner-city but also female, handicapped, and rural— have been hampered by inequitable access to computers and by widespread patterns of inequitable distribution" (p. 2). Neuman explained that beyond race and ethnicity, technology inequity is evident across gender, geography, and ability, among other demographic categories, and has great implications for student achievement.

The implementation of new technologies can either intensify or ameliorate existing inequities (Warschauer et al., 2004). For example, there are inherent biases in technology use as students from affluent backgrounds are more likely to be exposed to technology for critical thinking and project-based learning, while students from low-income backgrounds are more likely to be exposed to technology for rote skill development. Furthermore, the emphasis on rote skill development for those from low-income backgrounds often extends beyond the classroom into libraries and other educational contexts where students can access technology.

According to the Center for Great Public Schools (2008), "educators in senior high schools, particularly in urban areas, need greater technological support to help set up and use technology in their classrooms" (p. 2). Support rendered in these educational contexts must be actively and purposefully monitored to ensure equity. Likewise, technology training in these school settings must be unbiased and culturally competent due to the "insidious ways in which teachers and schools embed messages in curricula and pedagogies" (Gorski, 2008, p. 354). Unless great care is taken, implicit biases may impact decisions about which students are taught rote skills and which are taught to use technology for critical thinking and research. Moreover, when examining equitable technology use, schools tend to frame access to technology as a privilege and/or reward system. In this respect, our personal experiences bear witness to instances when access to technology was restricted because of students' poor behavior—even though the poor behavior was not related to the technology itself. Consequently, well-behaved students may be accorded greater access to technology through a merit-based system (Warschauer et al., 2004). Such in-school practices further expand the digital divide.

Preservice Teacher Technology Training

Lei (2009) suggested focusing on preservice teachers who are digital natives when technology training is conducted because they tend to be more tech savvy, having enjoyed access to technology themselves while in school. Digital natives grew up with technology as a part of their daily routines and are comfortable with accessing the World Wide Web via laptops, smartphones, tablets, and taking advantage of Internet cafes for connectivity. While digital natives are ideal for preservice teacher technology training, teacher preparation programs must consider that many novice educators lack extensive expertise in implementing classroom technology to enhance instruction (Lei, 2009). In fact, the Center for Great Public Schools (2008) noted that "although educators do get technology training, most do not feel prepared to use technology for instructional purposes, especially for individualized instruction" (p. 2).

This perceived lack of confidence in how to use technology for instructional purposes may influence both practicing and preservice teachers' efficacy when implementing technology in teaching and learning, leading to greater technology inequity in schools. Furthermore, as Anderson and Maninger (2007) asserted, "a critical issue in teacher education relates to how best to prepare preservice teachers to integrate technology into their future classrooms" (p. 151). This issue is critical when preservice teachers themselves may be subject to inequitable practices, only to have inequitable practices replicated when they become classroom teachers. Bullock (2004) posited that "the attitudes, fears, and experiences of preservice teachers [serve] as both an enabling and disabling factor" (p. 233), meaning preservice teachers' educational beliefs greatly influences their future technology teaching practices. To put it simply, preservice teachers' beliefs will have either a positive or negative impact on the success of their future students. From an equity perspective, negative teacher beliefs about classroom technology use can widen the technology gap between underserved students of color and their affluent, White peers. As a result, students with the most need may not receive the type of technology instruction that has the potential to put them on par with their affluent peers who are already at an advantage due to resources availability.

Pedagogical Considerations for Teacher Educators

From a pedagogical perspective, Lei (2009) declared that "to help preservice teachers integrate technology into teaching in meaningful ways, technology cannot be taught as a separate and independent domain" (p. 93). Instead, preservice teachers must be taught to incorporate instructional technology as a pedagogical tool. Such a shift could improve and generally increase teacher technology efficacy and eventually enhance their students' learning. Technology modeling and application, when infused in teacher education programs, can alter the thoughts, practices, and competencies of preservice teachers prior to their entry into classrooms (Fleming, Motamedi, & May, 2007). Pope, Hare, and Howard (2002) earlier asserted that the integration of technology in preservice teachers' methods classes had the potential to increase students' comfort and confidence levels regarding the incorporation of technology in their future classes. In the same vein, Ertmer et al. (2003) found that electronic modeling (in those days presented via CD-ROM), increased preservice teachers' ability to integrate technology.

In order to effectively assist preservice and practicing teachers in the integration of technology, certain strategies must be implemented. At the turn of the century, Ertmer (1999) recommended modeling, reflecting, and collaboration as specific strategies needed to circumvent the challenges faced by educators during technology integration. Those strategies are just as valid today, but additional strategies have been proposed more recently. For example, Brush and Saye (2009) outlined three different strategies, which were implemented with preservice social studies teachers learning to integrate technology into their future classrooms. These strategies were

1. Viewing, critiquing, and discussing authentic cases of social studies teachers utilizing various technology resources to implement inquiry-based learning activities in their classrooms;
2. Providing preservice social studies teachers with opportunities to explore innovative, emerging technologies and to integrate those technologies into rich learning activities within the context of their teacher education programs;
3. Providing preservice social studies teachers with opportunities to implement activities that effectively utilize technology in authentic classroom settings. (p. 47)

These strategies and others may facilitate teachers' understanding and application of technology, expanding the use of technology in K–12 classrooms and, in turn, facilitating their students' learning. For example, preservice teachers receiving adequate, culturally responsive computing (Scott, Sheridan, & Clark, 2015) training may in fact lessen the digital divide and begin to restore equity into school technology use. As preservice teachers receive culturally responsive methodological and pedagogical training, technology plays an important role in keeping the promise that every child will receive an appropriate education. The conditions under which teachers work and students learn must reflect a school culture that accords equitable access to appropriate technology-infused learning the highest priority.

METHOD

This study used a single-case study design (Yin, 2014). We chose a holistic design because it explored the global nature of technology use within a school- and community-based immersion program. The rationale for using a single case in this study is Yin's (2014) common case characterization. A common case is one in which "the objective is to capture the circumstances and conditions of an everyday situation" because of the insight that might be gained about "social processes related to some theoretical interest" (p. 52). With this understanding, it was our intent to capture how prospective preservice teachers utilized university technology to aid their underserved high school students in constructing projects focused on strengthening their students' math and English comprehension. In addition, context was extremely important to us when we considered the design of this study. Yin posited that "every type of [case study] design will include the desire to analyze contextual conditions in relation to the 'case'" (p. 50). Acknowledging the context within this single-unit case, we next provide a description for the Student Success Program (a pseudonym, as are all names in this chapter), as well as the urban school and community.

Student Success Program

Conceptually, the SS Program served as an urban teacher preparation initiative. It was a tutoring and mentoring initiative that prepared and connected college students with high school students to provide the high school students with experiences that supported their academic achievement in mathematics and English language arts, as well as their social skill development. With respect to their social skills development, the high school students participated in social events and activities in and around the university and greater city to serve their communities.

In designing the SS Program, the SS leadership team examined the best research in the field and concluded that to effectively tutor and mentor high school students, college students must complete seminar training sessions each semester to build five program competencies: (1) urban context within the U.S. education system, (2) pedagogy, (3) tutoring and mentoring in mathematics, English, financial literacy, life skills, and study skills, (4) participatory action research, and (5) integrating arts and technology into education. As a capstone assignment, college students completed an action research study and an art-focused technology project that furthered the high school students' academic and social progression. The participatory action research study served as a prerequisite to the art-focused technology project. After gaining insight about their high school students' academic and social challenges through the participatory action research study, the college students used this information to inform the direction of their art-focused technology project.

Participants

The SS Program initially operated as a pilot during its first 2 years. As a pilot, the SS Program involved 12 college students in the first cohort, each tutoring and mentoring two high school students. The college and high school student cohort remained relatively intact for the 2-year period. However, the program did experience a level of attrition, losing one college and six high school students over the course of 2 years. Additional participants were recruited, not exceeding the designated program design number (i.e., 12 college students and 24 high school students). Among the college student participants, there were 11 sophomores and one junior. There were also 11 female students and one male student. With regard to racial-ethnic composition, the majority of the college students were White, while there were two college students of color. Ten of the college students were liberal arts majors (e.g., English, psychology, political science, linguistics, etc.), one college student was a math and computer science major, and one college student was a neuroscience major. In this cohort, six college students wanted to pursue careers in education, while the other six college students wished to enter professions that were associated with education and social-justice.

The high school student group consisted of 24 ninth-grade students. Of this group, 15 were female and nine were male. The majority of the high school students were African American, with one Latinx participant and one White participant. The students ranged across the spectrum of academic proficiency and disciplinary profile.

School and Community Context

Revolutionary Prep is a Grade 6 through Grade 12 public school in a major, urban district. It is located in a working-class, historically Black,[2] urban community in the northeastern region of the United States. In addition, Revolutionary Prep is also located on the campus of a major state research university. Revolutionary Prep is a relatively small school that

Table 7.1.
Student Success Participant Demographic Information

College Student Participants			
Gender-Race	Black	White	
Male	0	1	
Female	2	9	
High School Student Participants			
Gender-Race	Black	White	Latinx
Male	9	0	0
Female	13	1	1

enrolls a mostly homogeneous student population of approximately 549 students. Ninety-one percent of students are African American, 4% are White, 3% are multiracial, 1% are Asian, 1% are Latino, and 1% Pacific Islander. Despite a majority student population of color, the school has a predominately White teaching faculty. It is also a Title I school with 89% of students qualifying for free or reduced-price lunch.

Revolutionary Prep needs additional funding to serve the social-emotional and academic needs of its students from low income backgrounds. Lacking these resources, Revolutionary Prep is characterized by a high incidence of student discipline referrals, chronic student absenteeism, and low proficiency standards on state standardized tests. State standardized test scores indicate that less than 50% of students met proficiency standards annually. To combat many of the out-of-school sociological issues that impede teaching and learning inside of the school, the school relies heavily on strong community organizations, such as the local district education council, to fight for more equitable resources, services, and outcomes for the students.

In terms of technology capacity, Revolutionary Prep's technology was outdated, restricted, and poorly aligned with students' learning needs. The single computer lab and Windows PC laptop computers were frequently locked. Furthermore, when students checked out Toshiba Netbook computers they were often not fully charged. Although students had limited access to computers, they did have the option to take a computer course, but many school computers were painfully slow because they were purchased in 2011. Given that Revolutionary Prep was unable to purchase new computers due to lack of funding, students contended with ineffective technology resources that did little to further the learning process.

Data Collection

During the course of 2 years, numerous data were collected from participants. Although participants spoke and wrote about numerous topics and experiences, for the purposes of this study, we only highlighted and analyzed data that were relevant to the technology component of the SS Program. The data that were pertinent to this study were the high school students' technology assessment forms, pre- and post-in-depth, semistructured interviews, and the college students' journal reflections. Information was gathered from different sources to ensure data triangulation (Denzin, 1978).

The high school students' technology assessment forms were collected prior to their entry into the program; prospective participants submitted this form with their admissions applications. The Student Success Technology Assessment inquired about students' access to technology. Table 7.2 shows the questions that were posed to students.

Individual and group student interviews were also conducted with college and high school students to gain additional insight about students' technology experience and use. The interview protocol for the high school students' interviews encompassed a broad range of questions (e.g., students' school experiences, future plans, community involvement, technology, etc.), but, as mentioned above, this study focused only on students' technology responses. These interviews lasted approximately 90 minutes and were later transcribed. In addition to their interviews, the college students' journal reflections on technology were also examined. Weekly journal reflections were approximately one-page entries that were designed to help participants debrief on, make sense of, or document observations during their most recent tutoring and mentoring experiences. Lastly, students' collaborative technology projects added additional insight. These projects focused on different areas: some addressed English and math comprehension, while others were developed to assist in high school students' social development. Table 7.3 provides information about the students' technology projects.

Data Analysis

Data were inductively analyzed in this study using conventional content analysis which "is generally used with a study design whose aim is to describe a phenomenon" (Hsieh & Shannon, 2005, p. 1279) and allows for a "subjective interpretation of the content of text data through the systematic classification process of coding and identifying themes or patterns" (Hsieh & Shannon, 2005, p. 1278). In the data analysis process, we initially

Table 7.2.
Student Success Technology Assessment

Name:	Grade:

Please circle all responses that apply

1. Do you have Internet access where you live?	Yes No
2. If you do not have Internet access where you live, where else could/do you access the Internet?	a. Library b. House of a relative or friend c. Other (Please write in a response) _____
3. On which of the following devices could/do you access the Internet?	a) Desktop computer b) Laptop computer c) Tablet d) Cell phone/smart phone e) Other (Please write in a response) _____
4. If you know, which operating systems are you familiar with?	a. Mac OS X b. Microsoft Windows c. Other (Please write in a response) _____ d. None of the above
5. Which of the following computer programs/ websites are you familiar with?	a. Google Drive b. Microsoft Office and Outlook c. Khan Academy d. Other (Please write in a response) _____ e. None of the above

Table 7.3.
Student Success Technology Projects

Groups	Focus	Description
A	math and social development	The Sky is the Limit website is a personal site where students can create blog posts and engage in math tutorials
B	English and math	The Lexx and Char show is an audio podcast devoted to discussions and raps about English and math concepts and historical events
C	English and social development	Theatrical webisodes in which students create and act out a plot framed around literary concepts
D	English and social development	Online journal, fictional short stories, and poetry
E	math	A computer program that prompts students to answer randomly generated algebra or geometry questions. The program keeps track of answers and gives a percentage score at the end, motivating students to continue to play for a higher score.
F	English and social development	An online interactive resume and portfolio

Note: Each group consisted of two college students and four high school students.

collated data that highlighted participants' technology experiences. Once this process concluded, we read the technology assessment forms, interview transcriptions, and journal reflections in their entirety and gained a holistic understanding of those data. Then we independently engaged in open coding (Corbin & Strauss, 2015), each of us constructing our own distinct codes. This word-by-word, line-by-line examination of texts produced a variety of codes, revealing distinct and interrelated concepts. We then came together and compared our findings, determining commonalities among our respective codes. Lastly, we combined similar codes and developed emerging categories which enabled us to distill salient themes. We discussed these themes until we reached consensus. We checked the validity of our findings through collegial debriefing and by persistent field observations conducted over a 2-year period (Lincoln & Guba, 1985).

FINDINGS

Through the technology arm of the SS Program, college and high school students co-constructed an art-focused technology project, demonstrating how a 2-year, school-university-community collaboration can positively influence the digital divide and the technology experiences of underserved students of color. As mentioned above, a participatory action research study served as a prerequisite to the art-focused technology project. After gaining insight about their high school students' academic and social challenges through the participatory action research study, the college students used this information to inform the direction of their co-constructed, art-focused technology project.

The student technology assessment forms were given to the high school students to gain an initial assessment beyond the school setting of students' access to and familiarity with digital technology. The student technology assessment forms and the high school student pre-interviews indicated that, while the majority of the high school students in the SS Program had Internet access, Internet use was often limited to cell phones. This limitation may explain why many of the high school students were not familiar with different operating systems such as Windows or Mac OS X. Aware of students' differing levels of proficiencies with technology, the college students used this information to inform their pedagogical practices and frame their instructional interaction with their students. For instance, rather than having the high school students work independently with technology (mainly laptop computers at the urban school site), students were placed in cooperative learning groups and had consistent guidance from their college mentors.

Prior to beginning the technology project with students, one college student reflected extensively in her journal about the importance of technology access and use for underserved, students of color attending an urban public school. Her thoughts captured many of the issues students face, despite the ubiquitous nature of technology. Stephany recounted:

> I think our technological project is very important for our mentees. Coming from a school that lacks the resources that wealthier school districts have, the students may not be exposed enough to technology. I think the technological project is especially important because some of the students at [Revolutionary Prep] may have the perception that technology's sole purpose is social and for fun, but they may not realize all the informative, educational purposes technology can have. I have noticed that most of the kids in the school have cell phones, but I am curious to know how many of the students can use Microsoft Office programs effectively. This is why I think exposing our mentees to the technology outside their cell phone is so vital. In the world we live in today, where technology is growing exponentially, technology skills are even more valuable than ever before. Being adept in technology is a vital skill to have, and hopefully through our technological project our mentees will grow in their technological skills.

Within the urban school setting, the college students initially observed how technology was substituted for adequate teaching. Amy recounted her experiences in a math classroom in which a teacher did not incorporate differentiated teaching instruction into his classroom lessons. Instead, he used computer videos from Khan Academy[3] to teach difficult math concepts. In this context, Amy described her role in cultivating a more equitable technology environment for her students. She wrote:

> Honestly, during class I think some of [the students] are bored and that is why they are dancing; they're on their phones. I play multiplication games with them using the available technology. [The teacher] usually allows it because he often does not teach. He just puts computers in front of them that displayed Khan Academy.

Overall, many of the high school students lacked fluency when operating computers. Students' lack of in-class technology instruction from classroom teachers and inexperience with hands-on applications surely hindered students' technology proficiency. Students' lack of confidence with technology was observed by all of the college students. For instance, when recounting an after-school tutoring session with her student in her journal, Stephanie recalled her student's diffident engagement with technology. She explained:

> On Thursday, [Luna] came after school. We finished her math homework really fast, and then I gave her freedom to do a story for the technology project. I was really proud that she spent so much time and wrote a lot, even though she writes very slowly on the computer.

As the college and high school students began the process of completing their technology projects, they encountered many impediments. In fact, completing the technology projects at the urban school site proved extremely challenging. Many of the prospective preservice teachers described this issue in their interviews. Delphine explained how through the process of completing the technology project, she gained a greater awareness about the lack of technology resources available to her students. She reflected that

> I think limited resources at [Revolutionary Prep] was the biggest. I think this technological project is how I realized the lack of resources that the students have at [Revolutionary Prep] in technology. The computers run really slowly is one thing. They aren't able to access as many platforms. In building a computer code we had to use javascript and python but neither one of them is supported by the school computers. We had to go to the university to access these software. That was my biggest challenge.

Adding to Delphine's experiences, Madison expressed in her interview that "the fact that the computers [didn't] run very well was really frustrating." She closed by saying, "if I could change anything, I would give [my students] computers they could use." Although limited access to technology was a new phenomenon for the college students, restricted computer use was the lived experiences of the high school students. In fact, "these computers are slow" was a common statement echoed in many of the high school student pre- and post-interviews.

Another common theme running through the high school student interviews was an awareness that technology was better at more affluent schools. Kendra, a ninth-grade student proclaimed, "I bet those White kids in the suburbs have new computers that run really, really fast." Overall, the high school students were extremely frustrated by their lack of access to adequate technology.

Although students at Revolutionary Prep did not have access to up-to-date technology, their school was located on the campus of a major, state research university where technology was accessible in every building. Within walking distance of Revolutionary Prep at the university site were Mac computers, projectors, iPads, audio and video recording devices, high-speed Internet, etc. Because of the urban school's close proximity to that major university, the college students, unable to make significant gains with the urban school site technology, brought their high school students to the university's computer lab in order to utilize available university technology. Sophia explained how having more than one working computer among her group was beneficial to student productivity and peer support. With adequate technology, the high school students could work side-by-side and the college students did not have to divide their time, engaging students in two different activities. She captured this scenario by saying, "when I had both students [at the university], they were both on their sites, editing and doing stuff together and asking each other questions." She went on to say that

we did a soundcloud podcast. They were able to upload podcasts and compose spoken word or songs that they made. That was more for creative expression but we [did] math songs too. I think everyone has heard the songs and we used audacity to edit the song.

Jamie recounted her own experiences utilizing university technology with her students. She stated "I took [Kendrick] to the university and it was really helpful. It was a different environment. It was not necessarily quieter, but he was more focused and did not have distractions and everything was fast."

The high school students also relished the improved technology experiences at the university computer lab compared to their school computer lab—which was often locked. As Jamal, an eleventh-grade student, concisely stated, "the computers at the university were fast. I wish our school had those types of computers. It didn't take me forever to get all my work done." Shana, another ninth-grade student, expressed how she felt more comfortable using technology because she worked with her preservice teacher over the course of 2 years. She explained,

> Nicole and I spent a lot of time working on our technology project, and we always used the computers at [the university]. She showed me the different computer features. I feel a lot more comfortable with computers. I even have my own YouTube channel now. It has twenty-six subscribers.

Nicole, Shana's mentor, also spoke of their process as they worked together to complete their technology project. She described how technology boosted Shana's confidence in writing, reflecting that

> we did a WordPress blog. It was used to track her goals, life learning and academic and social goals. We did a lot of cool stuff with it. Having auto-correct through the computer also really helped her because one of her academic weaknesses is spelling, so using that feature on the computer as a resource gave her more confidence to write because she could move faster, and she didn't get caught up on words.

Lastly, the high school students conveyed that, after gaining great exposure to technology, they were more open to and now considered the idea of going into a technology-related career field. Many of the students described a possible career in programming (specifically, creating video games) while other students expressed an interest in web design. These findings, though limited to one urban school, show how the actions of prospective preservice teachers over the course of 2 years can assist high school students gain greater knowledge of digital technology and enhance their confidence in their ability to use it.

DISCUSSION

In the SS Program, both the preservice teachers and the high school students benefitted from collaborative technology experiences: the preservice teachers gained greater proficiency in technological instruction and the high schools students expanded their technology knowledge-base. Engaging in technology learning experiences and using campus equipment, participants learned from each other. Furthermore, access to such technology experiences and opportunities has the potential to positively impact students' lives, and thus, their community.

While equity in digital technology remains an issue, especially in urban school settings (Kormos, 2018), the instructional practices of preservice and practicing teachers alike can enrich students' technology experiences. The findings of this study showed that the majority of the students in the SS Program had access to the Internet via their own personal cell phones, but few gained access through a home desktop or laptop computer. In addition, access to in-school technology was often restricted and, when available, was frustrating to use. The limited accessibility and the poor quality of access have the potential to impact students' opportunity to learn digital technology and hinders their ability to fully engage and achieve academic gains. This finding aligns with the Freeman et al. (2017) theme of "widespread use of technology does not translate into equal learner achievement" (p. 4). Although students may have been proficient in operating their cell phones, accessing the Internet solely via a mobile device limits the extent to which a student can learn to use digital technology. This was the situation at the beginning of the SS Program due to inequitable, in-school opportunities to utilize digital technologies. Such a suboptimal learning environment correlated the high school students' observed lack of confidence and low levels of fluency in operating academic-related technologies.

As Chen (2010) correctly asserted, "access does not mean only the availability of hardware and software but also the appropriate type of technology-related factors" (p. 3). Outlining factors that contributed to unsatisfactory learning outcomes, he provided recommendations for targeted training that is more instructionally centered and subject-area specific. His recommendations included the use of assistive technologies to support special populations and ensuring proper training of those charged with educating students. In the light of Chen's recommendations, it is apparent that good pedagogy necessitates the understanding that technology serves as an instructional tool which should be implemented to enrich, not substitute for effective teaching. Unfortunately, students in the SS Program often engaged with technology in isolation while at school. Such narrow participation with technology actually decreases its effectiveness. Indeed, there is no replacement for good teaching. In order to ensure that students progress academically and are adequately prepared for the digital world, preservice and practicing teachers must serve as "guides, mentors, and coaches to help [students] navigate projects, generate meaning, and develop lifelong learning habits" (Freeman et al., 2017, p. 4).

Simply providing K–12 students in underserved, urban school with technological tools is insufficient, especially when the technology is outdated. Encouraging effective practicing teacher behavior begins in teacher preparation programs. Preservice teachers must be taught technological pedagogical content knowledge (Angeli et al., 2016), as well as how to properly embed technology into their instructional practices while simultaneously meeting the needs of their students. Technology should not be viewed as an "optional add-on" to the curriculum, separate from and exclusive of instructional pedagogy (Bullock, 2004). While subject-driven technology training should be a part of every preservice teacher training program, it should also be taught through a culturally responsive, multicultural lens (Gay, 2010; Gorski, 2008; Villegas & Lucas, 2002) in order to be responsive to students' diverse learning needs. Culturally responsive computing instruction (Scott, Sheridan, & Clark, 2015) in teacher education may lessen the equity issue that continues to sustain the digital divide. This training would not only make teachers aware of technology inequities between schools and school districts, it would also help them identify within-school classroom practices that exacerbate the digital divide. Most importantly, in practicing culturally responsive computing, future educators can learn how to tailor their technology classroom instruction based on students' cultural background and experience.

Improved digital technology experiences begin with access to adequate hardware and responsive instruction that challenges students to engage with technology beyond surface-level comprehension. Almost 20 years ago, Baylor and Ritchie (2002) declared that "technology impact on higher-order thinking skills was predicted by teacher openness to change … and negatively influenced by the percentage of technology use, where students work alone" (p. 1). Subsequently, Gorski (2008) concurred that "students who are least likely to have access to higher-order instruction without these technologies are also least likely to have access to such instruction when these technologies are in play" (p. 355). In order to ensure equitable, technology enriched school environments, critical, higher-order thinking skills must be infused into classroom technology instruction because, in spite of the narrowing trends on race-based gaps presented in the 2017 NAEP assessment, there are still great academic disparities between different student groups. In fact, "the gap between high- and low-achieving students widened on a national math and science exam, a disparity that educators say is another sign that schools need to do more to lift the performance of their most challenged students" (Balingit, 2018, p. 1). The thoughtful integration of digital technology, tailored curricula, and responsive pedagogy has the potential to impact the equity divide in technology, as well as the potential to lessen the persistent achievement gaps between different racial/ethnic student groups.

CONCLUSION

While the presence of digital technology is pervasive in our society, we must take steps to make inroads into the inequitable practices and systemic structures that continue to impact the technology experiences of under-served students of color. Ensuring that all students have opportunities to learn and achieve with digital technologies is of paramount importance considering technology's integral role in daily life. Student's inculcation of a fundamental digital technology skill set is invaluable. That said, the effective pedagogical practices of preservice and practicing teachers when integrating digital technology into their lessons can improve the digital technology confidence and skills of students. To contribute to decreasing the digital divide and eventually ensuring its demise especially in under-resourced, educational contexts, programs similar to the SS Program that prioritize creative and intentional teaching must occur. The absence of such initiatives and facilitative pedagogies will surely sustain the status quo.

NOTES

1. A person born or brought up during the age of digital technology and there-fore familiar with computers and the Internet from an early age.
2. The authors use the term Black or African American interchangeably to identify persons from the African Diaspora.
3. Khan Academy is a nonprofit educational organization that provides a set of online tools that help educate students.

REFERENCES

Anderson, S. E., & Maninger, R. M. (2007). Preservice teachers' abilities, beliefs, and intentions regarding technology integration. *Journal of Educational Computing Research, 37*(2), 151–172.

Angeli, C., Voogt, J., Fluck, A., Webb, M., Cox, M., Malyn-Smith, J., & Zagami, J. (2016). A K–6 computational thinking curriculum framework: Implications for teacher knowledge. *Educational Technology & Society, 19*(3), 47–57.

Balingit, M. (2018). National math and reading remain constant, but dispari-ties emerge. *The Washington Post,* pp. 1–4. Retrieved from: https://www.washingtonpost.com/local/education/national-math-and-reading-scores-remain-constant-but-disparities-emerge/2018/04/09/4ad92714-3c0c-11e8-8d53-eba0ed2371cc_story.html?utm_term=.80cba88a610a

Baylor, A., & Ritchie, D. (2002). What factors facilitate teacher skill, teacher morale, and perceived student learning in technology-using classroom? *Computer & Education, 39*(1), 395–414.

Brush, T., & Saye, J. (2009). Strategies for preparing preservice social studies teach-ers to effectively integrate technology: Models and practices. *Contemporary Issues in Technology and Teacher Education, 9*(1), 46–59.

Bullock, D. (2004). Moving from theory to practice: An examination of the factors that preservice teachers encounter as the attempt to gain experience teaching with technology during field placement experiences. *Journal of Technology and Teacher Education, 12*(2), 211–237.

Center for Great Public Schools. (2008). *Technology in schools: The ongoing challenge of access, adequacy, and equity.* Retrieved from: https://www.nea.org/assets/docs/PB19_Technology08.pdf

Chen, R. J. (2010). Investigating models for preservice teachers' use of technology to support student-centered learning. *Computers & Education, 55*(1), 32−42. https://doi.org/10.1016/j.compedu.2009.11.015

Corbin, J., & Strauss, A. (2015). *Basics of qualitative research: Techniques and procedures for developing grounded theory* (4th ed.). Los Angeles, CA: SAGE.

Denzin, N. K. (1978). Triangulation: A case for methodological evaluation and combination. In N. K. Denzin (Ed.), *Sociological methods: A sourcebook* (2nd ed., pp. 339–357). New York, NY: McGraw-Hill.

Dolan, J. E. (2016). Splicing the divide: A review of research on the evolving digital divide among K–12 students. *Journal of Research on Technology in Education, 48*(1), 16–37. https://doi.org/10.1080/15391523.2015.1103147

Ertmer, P. A. (1999). Addressing first-and second-order barriers to change: Strategies for technology integration. *Educational Technology Research and Development, 47*(4), 47–61.

Ertmer, P. A., Conklin, D., Lewandowski, J., Osika, E., Selo, M., & Wignall, E. (2003). Increasing preservice teachers' capacity for technology integration through the use of electronic models. *Teacher Education Quarterly, 30*(1), 95–112. http://www.jstor.org/stable/23478427

Fleming, L., Motamedi, V., & May, L. (2007). Predicting preservice teacher competence in computer technology: Modeling and application in training environments. *Journal of Technology and Teacher Education, 15*(2), 207–231.

Freeman, A., Adams Becker, S., Cummins, M., Davis, A., & Hall Giesinger, C. (2017). *NMC/CoSN Horizon Report: 2017 K–12 Edition.* Austin, TX: The New Media Consortium.

Gay, G. (2010). *Culturally responsive teaching: Theory, research, and practice.* New York, NY: Teachers College Press.

Gorski, P. C. (2008). Insisting on digital equity: Reframing the dominant discourse on multicultural education and technology. *Urban Education, 44*(3), 348–364. https://doi.org/10.1177%2F0042085908318712

Hess, F. M., & Leal, D. L. (2001). A shrinking "digital divide"? The provision of classroom computers across urban school systems. *Social Science Quarterly, 82*(4), 765–778.

Hsieh, H.-F., & Shannon, S. E. (2005). Three approaches to qualitative content analysis. *Qualitative Health Research, 15*(9), 1277–1288. doi:10.1177/1049732305276687

Jackson, L. A., Zhao, Y., Kolenic, A., Fitzgerald, H. E., Harold, R., & Von Eye, A. (2008). Race, gender, and information technology use: The new digital divide. *Cyberpsychology & Behavior, 11*(4), 437–442. https://doi.org/10.1089/cpb.2007.0157

Kormos, E. M. (2018). The unseen digital divide: Urban, suburban, and rural teacher use and perceptions of web-based classroom technologies. *Computers in the Schools, 35*(1), 19–31. https://doi.org/10.1080/07380569.2018.1429168

Lei, J. (2009). Digital natives as preservice teachers: What technology preparation is needed? *Journal of Computing in Teacher Education, 25*(3), 87–97. https://doi.org/10.1080/10402454.2009.10784615

Lincoln, Y. S., & Guba, E. G. (1985). *Naturalistic inquiry.* Beverly Hills, CA: SAGE.

Mouza, C. (2008). Learning with laptops: Implementation and outcomes in an urban, underprivileged school. *Journal of Research on Technology in Education, 40*(4), 447–473. https://doi.org/10.1080/15391523.2008.10782516

Neuman, D. (1991). *Technology and equity.* Syracuse, NY: ERIC Clearinghouse on Informational Resources. (ERIC Document Reproduction Service No. ED 339 400).

Organization for Economic Co-operation and Development (OECD). (2002). *OECD information technology outlook.* Paris, France: Author. Retrieved from https://www.oecd.org/sti/ieconomy/1933354.pdf

Organization for Economic Co-operation and Development (OECD). (2004). *OECD information technology outlook.* Paris, France: Author. Retrieved from https://www.oecd.org/sti/ieconomy/37620123.pdf

Pope, M., Hare, D., & Howard, E. (2002). Technology integration: Closing the gap between what preservice teachers are taught to do and what they can do. *Journal of Technology and Teacher Education, 10*(2), 191–203.

Scott, K. A., Sheridan, K. M., & Clark, K. (2015). Culturally responsive computing: A theory revisited. *Learning, Media, and Technology, 40*(4), 412–436. https://doi.org/10.1080/17439884.2014.924966

Staples, A., Pugach, M. C., & Himes, D. (2005). Rethinking the technology integration challenge: Cases from three urban elementary schools. *Journal of Research on Technology in Education, 37*(3), 285–311. https://doi.org/10.1080/15391523.2005.10782438

Villegas, A. M., & Lucas, T. (2002). Preparing culturally responsive teachers: Rethinking the curriculum. *Journal of Teacher Education, 53*(1), 20–32.

Warschauer, M., Knobel, M., & Stone, L. (2004). Technology and equity in schooling:Deconstructing the digital divide. *Educational Policy Analysis Archives, 18*(4), 562–588. doi:10.1177/0895904804266469

Wilson, N. (2014). Interrogating the divide: A case study of student technology use in a one-to-one laptop school. In J. L. Polman, E. A. Kyza, D. K. O'Neill, I. Tabak, W. R. Penuel, A. S. Jurow, ... L. D'Amico. (Eds.). *Learning and becoming in practice: The International Conference of the Learning Sciences (ICLS) 2014 Proceedings Volume 1* (pp. 448–454). Boulder, CO: International Society for the Learning Sciences. Retrieved from https://www.isls.org/icls/2014/downloads/ICLS%202014%20Volume%201%20(PDF)-wCover.pdf

Yin, R. K. (2014). *Case study research: Design and methods* (5th ed.). Los Angeles, CA: SAGE.

CHAPTER 8

INTEGRATING DIGITAL TECHNOLOGY IN EDUCATION

A Tech Center in the U.S. Borderland Region

Lucia Chacon-Diaz
The Ohio State University

Susan Brown
New Mexico State University

For over 20 years, the partnership between Gadsden Independent School District (GISD) and the STEM Outreach Center (SOC) at New Mexico State University (NMSU) has opened avenues that increase the well-being of the communities within the school district. Although GISD has the best intention to serve their student population, the district faces the challenge of lacking resources due to a high-poverty rate. Therefore, the SOC at NMSU sought a collaboration with the GISD through the implementation of out-of-school-time (OST) programs. The SOC prepares teachers in the GISD to implement hands-on, inquiry, and problem-based learning activities during after school hours. The types of OST programs have expanded over the years and have evolved to adopt innovative digital tools that will promote student learning.

Integrating Digital Technology in Education:
School–University–Community Collaboration, pp. 173–197
Copyright © 2019 by Information Age Publishing
All rights of reproduction in any form reserved.

173

Because of the work that the SOC at NMSU has done over the years of serving the communities in New Mexico, the Department of Defense chose NMSU as the first university to house a Test and Evaluation Collaboration Hub (TECH) Center laboratory. The TECH Center provides attendees with ultimate experiential learning experiences through the usage of state-of-the-art technology. The SOC hopes that by providing students at GISD with such authentic active learning experiences, the students will develop higher order thinking skills, become motivated to pursue a college degree at NMSU, and join the STEM workforce. It is through this partnership that the SOC seeks to impact, not only the school district, but also the community, to improve the economic well-being of the Gadsden area.

The future economic well-being of the United States greatly depends on students with internationally competitive skills in mathematics, language arts, and science (Zheng, Stapleton, Henneberger, Woolley, & Maryland Longitudinal Data System Center, 2016). In the southern region of New Mexico, the majority of the students that the Gadsden Independent School District (GISD) serves are Hispanic, English language learners and have a low socioeconomic status (Children, Youth Families Department, 2014; New Mexico Public Education Department, 2016). Since the percentage of children living in poverty is twice the national average, GISD is considered one of the poorest districts in the nation. Thus, through the Science, Technology, Engineering, and Mathematics (STEM) Outreach Center at New Mexico State University (NMSU), students of GISD are offered out-of-school-time (OST) programs to increase their participation and achievement in the STEM fields. Examples of these programs include: SEMAA (Science, Engineering, Mathematics, and Aerospace Academy) and DiMA (Digital Media Academy), COUNT (Creating Opportunities Using Numerical Thinking), Science Brain Battle, Healthy Choices, and Reader's Theater.

The partnership between a university located in the southwest region of the United States and a school district with poor resources impacts the community of the school district. Being born into poverty was not a choice for the students in GISD. Students in GISD falling below the poverty line deserve the opportunity to learn (McDonnell, 1995), just like any child. For a child to be able to succeed in school, he/she must first be exposed to novel learning experiences supported by brain theory and learning research (Bransford, Brown, Cocking & National Academy of Sciences/National Research Council, 2000). Thus, the STEM Outreach Center (SOC) seeks to provide these families with an ounce of hope for a better future by implementing OST programs. The SOC has offered their outreach services to GISD for over 20 years. This partnership enables the flourishing of new academic avenues for teachers and their students.

In the *New Media Consortium and the Consortium for School Networking (NMC/CoSN) Horizon Report: 2017 K–12 Edition*, Freeman, Adams Becker, Cummins, Davis, and Hall Giesinger (2017) address the crucial role of technology in today's educational settings. The adoption of technology in the classroom in rural high-poverty middle schools positively impacts minority students, by increasing their academic performance in math and science (Blanchard, LePrevost, Tolin, & Gutierrez, 2016). However, this leads to the question of how many K–12 students living in rural high-poverty areas actually have the support from their district to provide them with technological resources. The establishment of OST in GISD provides teachers and their students with digital tools to support student learning and the development of their critical thinking skills.

This chapter addresses how the partnership was established and the trajectory it has undertaken until the most recent event which was the opening of the TECH Center at NMSU in the fall of 2017. The first sections of the chapter provide the national and New Mexico scope of STEM education and workforce. Then, a history of the partnership will be addressed, followed by a brief description of the current OST programs. Next, the positive impact of the collaboration through a descriptive analysis of GISD student performance will be described. Finally, the emergence of the TECH Center will be discussed, as well as the lessons learned throughout the years of partnering with GISD.

National Scope of STEM Education and Workforce

According to Darling-Hammond, Zielezinski, and Goldman (2014), every 29 seconds, one high school student in the U.S. drops out. The U.S. Department of Education (2015) stated that "The United States is falling behind internationally, ranking 29th in math and 22nd in science among industrialized nations" (The Need section, para. 3). Furthermore, the U.S. Department of Education has expressed concern that the number of students pursuing degrees in STEM remains low. Thus, the U.S. Department of Education established a national educational priority the increase in numbers of students' pursuing a career in the STEM fields. Subsequently, although there has been a slight increase of STEM degrees recently, there are still fewer workers in STEM occupations (Noonan, 2017).

Xue and Larson (2015) from the Bureau of Labor Statistics (United States Department of Labor) reported on the shortages of the STEM labor market with respect to government sectors and private industries. Citing Salzman and colleagues, Xue and Larson asserted that "for every two students graduating with a U.S. STEM degree, only one is employed in STEM" (para. 6). Further, Fayer, Lacey, and Watson (2017) stated that "there were nearly 8.6 million STEM jobs in May 2015, representing 6.2 % of U.S. employment. Computer occupations made up nearly 45 percent of STEM employment, and engineers made up an additional 19 percent" (p. 2). Noonan (2017) affirmed that "STEM occupations are projected to grow by 8.9 percent from 2014 to 2024, compared to 6.4 percent growth for non-STEM occupations" (p. 2). In other words, the STEM workforce is set to increase at a higher rate compared to the non-STEM workforce. Nonetheless, STEM areas, such as nuclear and electrical engineering, have a shortage of workers. The "competitiveness, economic growth, and overall standard of living" (p. 11) of the United States is greatly impacted by the STEM workforce. The issue of shortages in the STEM workforce may limit the prosperity of current and future generations. Since technology influences a great portion of our daily lives, we must find ways to maintain the U.S. as a viable international competitor.

New Mexico Scope of STEM Education and Workforce

A State Highlights Report published by *Education Week* (2018) considered all states' Chance-for-Success Index (the contribution of education to an individual's positive outcomes over his or her lifetime), their K–12 Achievement Index (18 measures related to reading and math achievement, high school graduation rates, and advanced placement outcomes), and their school finance analysis (school spending patterns and equitable distribution of funds across the state). The report concluded that New Mexico earned an overall score of 65.9% and a grade of D, placing it in 50th position. Based on this conclusion, the state of New Mexico has great need of federally funded programs that will improve the quality of K–12 education. In terms of STEM education, New Mexico ranks below the national average in female students' interest in pursuing a degree in STEM (Alliance for Science & Technology Research in America, 2018). Although the number of students graduating with a degree in STEM (in New Mexico) has increased over the years, the rate at which non-STEM majors graduate remains higher (Legislative Finance Committee Program Evaluation Unit, 2016). Also, "STEM graduates from out of state are far more likely to receive advanced degrees and far less likely to be employed in New Mexico" (Legislative Finance Committee Program Evaluation Unit, 2016, p. 12). Similarly, the Legislative Finance Committee Program Evaluation Unit (2016) speculated that, "New Mexico high tech industries prefer to hire STEM graduates from other states" (p. 11). Thus, there is a need to better prepare New Mexican STEM graduates for the high-tech workforce.

According to the New Mexico Department of Workforce Solutions (2016),

> over one-third of New Mexico's workers are employed in office/administrative support, sales, or food preparation and serving occupations. Science, technology, engineering, and mathematics (STEM) occupations, along with management occupations, were often the highest paying in New Mexico and other states. (p. 5)

Undoubtedly, job opportunities in STEM would greatly impact the economic well-being of New Mexico. However, the Legislative Finance Committee Program Evaluation Unit (2016) reports that, "employment in the high-tech industry has increased 22 percent nationwide since 2000, while New Mexico high-tech employment has declined 30 percent, from 25 thousand to 18 thousand workers, 23 percentage points lower than any other state" (p. 5). Therefore, measures must be undertaken within the state in order to increase high-tech employment for New Mexicans.

GISD CONTEXT

According to the Statewide Town Hall (2016), rural areas require the greatest attention because they occupy most of the New Mexico region. GISD is as large as Rhode Island. The New Mexico Public Education Department (2017) reports that 100% of the students in the GISD are economically disadvantaged, compared to the 74.1% overall rate of economically disadvantaged students statewide. In regard to gender, students in GISD are 48.3% female and 51.7% male. In terms of ethnicity, GISD has 96.9% Hispanic, 2.4% Caucasians 0.5% African American, 0.1% Asian American, and 0.1% American Indian students. Students with disabilities in GISD are 15.1% of the total student population, and 34.2% are English language learners.

History of the GISD-SOC Partnership

The partnership with GISD that is the focus of this chapter began in 1990. The partnership started when the current director of the STEM Outreach Center, Dr. Susan Brown, wrote a grant to the National Aeronautics and Space Administration (NASA). Through the NASA grant, the STEM Outreach Center established one of its core programs, SEMAA: Science, Engineering, Mathematics, and Aerospace Academy. SEMAA is an out-of-school time (OST) program with the main purpose to teach K–Grade

8 students about aerospace. In the program, students were involved in hands-on projects which they could take back home after their completion. The NASA grant allowed the STEM Outreach Center to build a laboratory to implement SEMAA. Since there was no room to establish the laboratory at the College of Education, the laboratory was built at the College of Engineering at NMSU.

Through the collaboration with NASA, the STEM Outreach Center (SOC) was thrilled to participate in a NASA nationwide program in which children would go to Houston, TX for a Saturday morning NASA program. However, parents living in GISD had few resources and were unable to pay for gas to travel to Houston, TX. Therefore, the STEM Outreach Center decided to design their own implementation plan to fit the GISD context. The SOC began to provide all the resources for the students and worked with the children during after school hours. Over the following years, other sources of funding have been obtained to continue providing GISD with STEM learning opportunities. Each source of funding is designated to a set of schools and a set of programs. For example, the 21st Century Community Learning Center Funding supports the OST programs in GISD school sites. The schools served in GISD through this grant are the following elementary schools: Anthony, Chaparral, Desert Trail, Desert View, Loma Linda, Riverside, Santa Teresa, Sunrise, and Vado.

Relationships. Since New Mexico State University is a land grant university, we honor our commitment to helping our community and our state. Partnering with GISD has been a joyful experience because they are whole heartedly welcoming. The first meeting to initiate the partnership was with the superintendent, and then the associate superintendent. During the meetings, the SOC director noticed how excited both the superintendent and the associate superintendent were to begin the partnership. The district wanted to provide their student population with as many opportunities as possible. Since the grant did not cover money for buses, the district provided buses to transport the participants home after the OST program. At the beginning of the partnership, the SOC began working with approximately half of the schools, and then increased the number of schools. The SOC currently serves 16 out of 24 schools in GISD, and there are plans to expand the number of schools it serves.

New Mexico is considered a relationship state, which facilitates our partnership with GISD. Members from the SOC attend every school with humility by recognizing that we are guests in the school. We ask staff members at the schools to refer to us by our first names, to establish a professional but approachable relationship. Partnering with GISD has been a learning experience for the SOC, as we learn that most students' parents are scared of school. Most of the parents shared that they do not have a Grade 8 education themselves, but they want better for their children.

When SEMAA started, we conducted family festivals and brought in guest speakers. Among those who attended the family festivals were cousins and aunts/uncles. During such festivals, the children presented their final hands-on project to their families.

Evaluation. To evaluate the SEMAA program, the SOC conducted surveys of both the students and their parents. The SOC honors the first language of the individuals that we are serving, so our participants can select a survey written in Spanish or English. In past surveys, parents expressed their desire to be good parents so that their children could pursue a professional career and be freed from the necessity to work in the fields to earn a living. Parents expressed how much they wanted their children to graduate from high school. Therefore, in early 2000, the SEMAA curriculum was expanded so each grade level could have its own curriculum.

The SOC in context. Also arising from the SEMAA surveys, SOC began to sponsor family festivals which were hosted monthly at specific schools with approximately 250 attendees per festival. Then, in 2001, SOC staff members noticed that the number of parents attending the family festivals drastically decreased. It did not take long for the staff to realize that multiple border patrol trucks were parked outside the schools that hosted the family festivals. The parents that attended had to be creative and park at other nearby schools to evade the border patrol trucks and avoid potentially being deported. To address this issue, the director of the SOC contacted border patrol to ask that officers refrain from parking at the host school during the family festivals. Fortunately, the border patrol officers complied, and the family festivals continued to welcome GISD family members.

Once the students participating in the OST programs started Grades 5, 6, or 7, they were brought to the engineering laboratory. The students were looking forward to coming to the university for OST, where one of the things they learned was that the university is not a scary place. Some students were initially intimidated by the thought of coming to the university since their parents did not attend college. (As part of the SOC's service to the community, it implemented programs to help parents fill out college applications and student loan forms for their teenagers.)

As part of the SEMAA program, the SOC asked astronomy students at NMSU to set up telescopes so students and their families could see the moon through them. During the event, the director of the SOC noticed an older lady sobbing. The SOC director approached her and asked "*¿qué pasa?*" (what's wrong?). The older lady responded that she never thought she would see the moon up-close in her lifetime. She also mentioned how she was overjoyed that her grandchildren were going to college—mistaking her grandchildren's participation in the OST program as the equivalent of attending college courses.

On another occasion, as the SOC director was walking out of the school during an OST session, she saw a man and a woman whom she presumed to be a husband and a wife looking through the window from outside a classroom. The director approached the couple and asked them if they needed any assistance. Through the tears in the woman's eyes, the couple expressed how thrilled they were because their child was playing and learning with NASA instruments. The parents had a sense of pride to see their child in a room filled with NASA logos and devices.

Grant activity. It is such events that motivate the director of the SOC to continue applying for funding, and she is clearly adept at doing so since over the course of 19 years she has been awarded more than $1 million dollars per year on average for a total of $23 million. The SOC director strongly encourages the program specialists working at the SOC to visit the schools that the SOC is serving, to be approachable, and to establish relationships with the school and the community.

The motive behind partnering with GISD was the fact that it is a district in need. Before starting writing for the first grant to NASA in 1990, the SOC director knew about the benefits of out-of-school-time programs. She was aware of the problems faced by the community. One particularly stark reality is that, in the late 1980s and early 1990s, children in GISD who were between 11 and 13 years old were being used by drug cartels to carry drugs. These children were called "mules," and the cartels would take advantage of them to transport drugs in exchange for a certain amount of money. Because most families in the GISD community were desperately poor, the young children would do as the drug cartels asked to make some money to help support their families. We have not heard about young children being "mules" since the SOC started implementing out-of-school-time programs in GISD.

The partnership with GISD is deeply rooted in helping the community by aiding students to succeed in school, especially in the STEM areas. The SOC seeks to impact current and future generations through STEM teaching and learning. The partnership provides students with a safe out-of-school-time environment where they can explore and acquire STEM knowledge and abilities. Most of the students that the SOC serves have not developed their STEM identity. By being exposed to STEM learning experiences, students become aware of the possibility of pursuing a college degree in STEM. As part of the partnership with NMSU, the Doña Ana Community College in Las Cruces, NM, helps GISD students with a pathway to the community college. From our experience, students attending the SOC OST programs are less likely to join gangs, get pregnant, or be involved with taking illegal drugs—in addition to being better behaved in school. Participating in OST programs gives students a sense of purpose as they continue on their educational journey.

Current (OST) Programs

Rural areas such as GISD have limited resources with which to enrich students' knowledge in both STEM and language arts. Since parents face the challenge of not being able to afford after school care, the SOC at NMSU offers their programs free-of-charge. For over four years, these programs have served the purposes of (a) increasing interest, engagement and understanding of STEM fields; (b) increasing reading skills; (c) providing experiences with art and music; (d) promoting lifelong healthy habits; (e) increasing parental/family involvement; (f) offering professional development for the participating teachers; and (g) increasing school attendance and promoting positive social behaviors. Additionally, the SOC programs have been designed to provide active learning environments that promote real world application of subject content and increase digital literacy, in addition to being aligned with relevant research in the neuroscience of learning (Bransford, Brown, Cocking & National Academy of Sciences, 2000; Bruce & Casey, 2012).

The OST programs at the SOC incorporate interdisciplinary STEM teaching and learning approaches. As previously mentioned, the first OST program implemented at GISD was SEMAA: Science, Engineering, Mathematics, and Aerospace Academy. SEMAA, as one of the core programs of SOC, has served 34,000 K–12 students to date and continues to attract participants. The goals of SEMAA involve serving historically underrepresented students through STEM activities to improve student academic achievement and to increase their participation in STEM. This OST program also engages parents and the communities of the participating students through family festivals.

Digital Media Academy (DiMA) OST program fosters digital literacy in GISD elementary and middle school students (Grades 4 through 8) through hands-on activities that incorporate digital media and educational technologies. Students enrolled in DiMA learn STEM topics by using innovative technology (e.g., film equipment and software) as a tool to collect and analyze data, and to report their findings. DiMA allows students to merge science and technology with their everyday lives in their own communities. This interdisciplinary OST program teaches students at a young age how to use technology as a learning tool, in a safe and wise manner.

Project GUTS-Growing Up Thinking Scientifically is an OST program for middle school students. Through problem solving and mathematical thinking, students engage in computer programing activities to enhance their computational thinking skills. Students are exposed to real-world problems that require them to design computer models and analyze their collected data. At the end of the program, students share their final projects with other participating students in the program in southern New Mexico.

COUNT (Creating Opportunities Using Numerical Thinking) is an OST program for kindergarten through Grade 5 students. COUNT merges art with mathematics to increase student interest, skills, and confidence in math. Through COUNT, students learn to recognize how math is encountered in their everyday lives. Students implement mathematical formulas and skills to create and evaluate their art projects. Science Brain Battle and Reader's Theater are other noteworthy OST programs that encourage students to increase their STEM knowledge and skills. Science Brain Battle is a competition among middle schools. Groups of students complete a set of challenges by putting into practice their scientific, mathematical, and problem-solving skills. Reader's Theater is oriented to kindergarten through Grade 6 students, implementing a literacy curriculum that encourages a hands-on and multisensory approach to learning. Students enact a children's story in a play with the purpose of developing literacy and vocabulary skills. STEM lessons are merged within the lessons according to the theme and content of the stories.

NMC/CoSN and OST. Addressing the needs highlighted in the NMC/CoSN Horizon Report: 2017 K-12 Edition (Freeman et al., 2017) involves innovating K-12 education through the use of technology. The OST programs align with various points of Freeman et al. (2017), primarily with the premise that "learners are creators" (p. 4). This is due to the fact that the SOC offers programs that incorporate "hands-on experiences that place learners in the role of creators" (Freeman et al., 2017, p. 40). At the SOC, students have the opportunity to use digital tools to code and build models that will serve a specific purpose according to the problem-based instruction applied during the program's session.

In addition, all the SOC programs previously described incorporate inquiry-based learning, project-based learning, and follow a student-centered teaching approach. Students engage in their learning processes through the use of novel technological resources offered at the SOC, thus creating an environment that facilitates authentic learning. This accords with the Freeman et al. (2017) definition of authentic learning as the "intersection of experiential learning and real-world experiences, where students are actively involved in the learning process and find ways to relate the topic and skills back to their own lives" (p. 26). Further, the SOC programs foster critical and complex thinking in students without overwhelming them with textbook information; rather, students have the opportunity to be active in their learning process through hands-on experiences. Through such experiences, students are free to make mistakes without facing penalty so that they are able to view every failure as a learning opportunity.

Positive Economic Impact of Collaboration

In 2018, in-depth economic research was conducted by Delgado—a doctoral candidate from NMSU's Economic Development program (Chavez, 2018). According to Chavez (2018), Delgado found that the SOC has saved the state of New Mexico close to $5 million in childcare at locations where the OST programs are being implemented. Another noteworthy finding was that the SOC has generated an economic output in New Mexico of $15 million since 2009 (Chavez, 2018). During the fiscal period July 1, 2015 through June 30, 2016, an agreement reached between GISD and the regents of NMSU allocated $104,745 for the following SOC programs: SEMAA, Scientifically Connected Communities, and Project GUTS- Growing Up Thinking Scientifically.

Positive Academic Impact of Collaboration

In 2010, an external evaluation of math and science learning in GISD was conducted. The evaluation results confirmed that students participating in SEMAA for 3 or more years scored significantly higher than students who had not attended SEMAA. Test scores for those students attending OST programs at GISD are almost as high as the scores of students in Los Alamos, NM. GISD attributed the high student achievement scores to the OST programs.

Krishnamurthi, Ballard, and Noam (2014) asserted that "development of a science or STEM identity involves multiple pieces: getting young people interested in STEM topics and professions; developing competence and a sense of confidence; and getting youth to envision themselves as contributors and participants in this enterprise" (p. 8). The partnership seeks to develop participants' STEM identity through the implementation of OST student-centered STEM programs. In a similar way, students in GISD should have the opportunity to become confident in their STEM skills, as well as develop a positive attitude towards school. Ball, Huang, Rikard, and Cotten (2017), stated that "when students believe that they can succeed in academia and when they value academic activities, they are more likely to express positive STEM attitudes" (p. 19). Student behavior in the classroom has been reported to improve in students participating in out-of-school-time programs (Krishnamurthi, Ballard, & Noam, 2014). Teachers at GISD report that student behavior has changed as a result of their students participating in the programs.

Descriptive analysis. Approximately 1,583 elementary students were served in GISD schools from fall 2017 to spring 2018. From the 1,583 participating students, grades for 1,261 were reported for English/reading and math for the fourth quartile of the 2017–2018 school year. The main reasons why some grades for English/reading and math were not reported vary, but one of the reasons was that kindergarten and Grade 1 students do not receive a letter grade from their teachers. Another reason was the if a student has an Individualized Education Program (IEP), different benchmarks apply. Lastly, students who only recently arrived in the U.S. do not qualify for reading assessments, and 13.7% of the student population in GISD fall into the recently arrived category. In order to enroll in the SOC out-of-school-time programs, students are not required to reveal whether they are recently arrivals.

OST programs have been found to improve academic performance and aid in closing the achievement gap between low and high socioeconomic students in elementary schools (Auger, Pierce, & Vandell, 2013; Krishnamurthi, Ballard, & Noam, 2014). Table 8.1 summarizes a descriptive analysis that compares the final letter grade of the whole school (for the 2016–1017 school year—the most recent data available) as reported by the New Mexico Public Education Department (2017), with the aggregated fourth quartile English/reading and math grades of students participating in the SOC OST programs in GISD (for the 2017–2018 school year; since the reality is that school letter grades seldom change from one year to the next, the previous year's whole school data is adequate for the purpose of this comparison). For this descriptive analysis, the English and math grades were aggregated by letter grades for each school, and calculated as percentages. Letter grades of SOC participating students for Anthony Elementary, Desert Trail Elementary, Riverside Elementary, and Santa Teresa Elementary, were similar to the average of the school grade. Most remarkable, the average letter grade for Sunrise Elementary was a D. However, close to half of the grades of students participating in the OST programs earned B grades (47.5%); only 7.5% of grades reported were D. Similarly, the grade average for both Chaparral Elementary and Loma Linda Elementary was C. Nonetheless, students participating in the SOC programs at Chaparral scored mostly B grades (34.1%), and C grades only accounted for (26.9%). Students participating in the SOC programs at Loma Linda scored more A grades (35.8%) and B grades (36.4%) than C grades (19.3%). Likewise, students participating in the OST programs in Vado Elementary received mostly A grades (38.5%) and B grades (28%), compared to the school average (C).

As previously noted, GISD faces the challenge of low student academic performance (New Mexico Public Education Department, 2016). However, through the OST programs, students increase their interest and knowledge in STEM and language arts, potentially leading them to achieve better grades in school. Data from the SOC programs indicate that 95% of students return to enroll in them. Also, parents claim high satisfaction

Table 8.1.

Comparison Between the Average School Achievement and Students' Achievement

		GISD Students' School Grade Aggregate for English/Reading and Math				
	Average Letter Grade for the School Year (2016–	4th Quarter (2017–2018) English/Reading and Math Grades for Students Participating in Out-of-School Time (OST) Programs				
Elementary School	2017)	A	B	C	D	F
Anthony*	B	51 (14.5%)	103 (29.3%)	97 (27.5%)	64 (18.2%)	37 (10.5%)
Chaparral†	C	46 (17.4%)	90 (34.1%)	71 (26.9%)	40 (15.2%)	17 (6.4%)
Desert Trail*	B	57 (21.4%)	93 (35%)	80 (30.1%)	28 (10.5%)	8 (3%)
Desert View	A	81 (26.8%)	119 (39.4%)	71 (23.5%)	21 (7%)	10 (3.3%)
Loma Linda†	C	63 (35.8%)	64 (36.4%)	34 (19.3%)	10 (5.7%)	5 (2.8%)
Riverside*	B	98 (26.5%)	136 (36.8%)	80 (21.6%)	46 (12.4%)	10 (2.7%)
Santa Teresa*	B	102(34.3%)	103 (34.8%)	64 (21.5%)	22 (7.4%)	6 (2%)
Sunrise †	D	48 (17.3%)	132 (47.5%)	77 (27.7%)	21 (7.5%)	0 (0%)
Vado †	C	84 (38.5%)	61 (28%)	51 (23.4%)	21 (9.6%)	1 (0.5%)

Note: Based on the percentages of the aggregated grades for students participating in the OST programs, OST students performed similarly to or better than the average whole-school grade.

*OST students performed similarly to the average of the whole-school grade.

†OST student performance was higher than the average grade for the whole school.

and affirm that their children talk about science during dinner. Informal education enhances the development of an academic-related culture within the household and, we believe that eventually it expands to the community.

One of the most recent accomplishments of the SOC at NMSU is the opening of the Tech Center in the fall of 2017. NMSU is the first university in the U.S. to house a Department of Defense education lab. The Tech Center seeks to provide students visiting the lab with numerous digital technology experiences, such as Aviation Simulators, that are designed to promote critical thinking skills to solve real-world problems. Beyond the GISD students, students in the south-western region of New Mexico and south-eastern Texas, are welcomed to the Tech Center during their after-school hours. The following section describes the emergence of the TECH Center and its significance to the borderland region of the United States.

A TECH CENTER IN THE BORDERLAND REGION

The SOC gives access to innovative technological resources to underrepresented students. Technology interventions have been found to positively impact young students' learning and attitudes towards STEM (Nugent, Barker, Grandgenett & Adamchuk, 2010), leading to the implementation of digital instructional strategies in middle school and high school classrooms across the United States (McKnight et al., 2016; Yarbro, McKnight, Elliot, Kurz, & Wardlow, 2016). However, technology literacy differences are associated with differences in socioeconomic status and ethnicity (Ritzhaupt, Liu, Dawson, & Barron, 2013). Making it possible for students who live in rural areas with scarce resources to have digital technology educational experiences is essential in the growth of the technology job market. The incorporation of digital technologies in education in a low-income rural region of the U.S. has the potential to improve the economy, not just for the served region, but for the whole nation.

According to Darling-Hammond, Zielezinski, and Goldman (2014), technology supports the learning of low achieving students when the technology is used to manipulate data and express ideas, as it is in the teaching approaches that are utilized within the SOC OST programs. Giving students the opportunity to interact and engage with technology has been shown to benefit students at-risk (Darling-Hammond, Zielezinski, & Goldman, 2014). However, limited access to technology in an area correlates with lower socioeconomic status and higher percentages of people of color. According to Darling-Hammond et al. (2014), in high-poverty areas, 3% of teachers indicated that their students have access to digital tools to complete assignments at home, whereas, by contrast, 52% of teachers in wealthier schools indicated that their students have access to a technological device to help them complete their assignments at home.

Lack of technology access increases technology-related stress and anxiety which may then negatively impact the likelihood that students will pursue a STEM career (Ball, Huang, Rikard & Cotton, 2017). Thus, there is a potential relationship between inequitable access to digital tools and attitudes towards STEM. Moreover, limited access to technology devices results in teachers not incorporating technology in their lessons (Dalal, Archambault & Shelton, 2017); such cases are common in high poverty areas. However, giving teachers the option of scheduling a session at the TECH Center during afterschool hours enriches their curriculum.

The Origin of the TECH Center

The decision to install the TECH Center was negotiated via a series of conversations with Congressman Pearce's office (representative of the

2nd District of New Mexico) and with NMSU President, Dr. Carruthers. The latter was the former governor of New Mexico and currently serves as president of NMSU. During the summer and fall of 2016, the interested parties had a couple of meetings and a formal presentation. The Test Resource Management Center (TRMC) within the Department of Defense (DoD), held a briefing led by Mr. Rob Heilman and Ms. Denise De La Cruz (from TRMC-DoD) for the NMSU representatives on the three legs of the DoD STEM initiatives: Test & Evaluation/Science & Technology (T&E/S&T) Research, Internships, and Outreach (TECH Center). In September 2016, the DoD T&E/S&T Principal Deputy Director flew in to meet Chancellor Carruthers and confirmed the decision to fund the creation of a TECH Center.

Department of Defense funding made it possible for the NMSU College of Education to turn ordinary classrooms into novel learning environments that engage students in a complex Science, Technology, Engineering, and Math (STEM) curriculum. As highlighted above, this is the first time that a TECH Center has been installed at a higher education institution. Thanks to the support from the NMSU administration as well as the extensive STEM Outreach programs, NMSU was chosen for the site for this new TECH Center. The installation of the TECH Center was completed on August 1, 2017, and, during this fall, students have engaged in new and challenging missions that promoted team work and role playing as they solved "real-world" problems.

The TECH Center and Its Role

The Test and Evaluation Collaboration Hub (TECH) Center houses a set of technologies that create a new immersive learning environment for STEM students of every age, by embedding learning objectives in an entertaining narrative and enhancing the story with virtual world game play and simulation technology. The TECH Center is configured with computer-based aviation platforms and simulators that engage students in a variety of missions that require critical thinking in applying core math and science competencies ranging from time-speed-distance calculations to advanced physics. The Center currently focuses on test and evaluation (T&E) scenarios and aviation, but it is envisaged that the center will evolve over time to include other areas (e.g., hypersonics, autonomy, and robotics).

The TECH Center is available to Grade 5 through Grade 12 students in the many NMSU STEM outreach programs on offer, as well as to NMSU undergraduates and to preservice and in-service teachers for professional development. Additionally, the TECH Center is being utilized to support White Sands Missile Range's (WSMR) professional development training.

STEM training requires the development of critical thinking and technical skills (Noonan, 2017). The problem-solving strategies implemented in the aviation simulators can potentially increase students' critical thinking skills and, thus, increase their academic achievement.

At NMSU, the SOC is committed to assisting students' development of their STEM fluency, regardless of their socioeconomic background. The TECH Center at NMSU encourages students to increase their STEM knowledge and use their imagination to solve realistic problems. The TECH Center provides a digital learning ecosystem to GISD students in which their cognitive, affective, and behavioral domains interact to facilitate their learning, confidence, and higher order thinking skills. Students have the opportunity to use state-of-the-art technology for an ultimate STEM learning experience. It is our goal that the TECH Center should address the current needs of the nation and the state of New Mexico by strengthening the STEM workforce and enhancing STEM education.

For example, the TECH Center houses aviation simulators that facilitate the implementation of geospatial technological practices in order to solve real-world problems. Nugent et al. (2017) have shown the positive impact of geospatial technology approaches on students' attitudes towards STEM. Interactive programs are reported to positively impact students with low-socioeconomic backgrounds (Darling-Hammond et al., 2014) as students are provided with the opportunity to explore and use the technology in a variety of ways to learn. To illustrate the interactive nature of the learning environment, the aviation simulator provides students with feedback pertinent to the problem that the students are given to solve. For example, students may be required to program and plan a flight, and, if they do not select the appropriate amount of fuel to complete their flight, their plane will crash.

Students are provided the opportunity to visit the TECH Center during OST hours and develop their digital literacy. The TECH Center "hooks" students with technology by exposing students to real-world problems (e.g., putting out fires in the Everglades, rescuing baby elephants from wildlife hunters, etc.). Because students at GISD are from low-income families and have limited resources to travel, the TECH Center incorporates geography within the simulators so students can explore different worldviews.

Computer-related occupations are growing at a fast rate, and the demand for workers with a degree in computer-related areas has been increasing (Fayer et al., 2017). Hopefully the TECH Center will address this crucial point by encouraging students to develop high-tech skills, and pursue computer-related degrees and occupations. Citing both Georgetown University and American Council on Education studies, a background report prepared for the 2016 Statewide Town Hall (2016) declared that "an estimated 80 percent of all jobs in the next decade will require STEM skills, and most of those positions will need some level of post-secondary training" (p. 24). This same document addressed the interrelationships among low job creation, inadequate education, and the high poverty rate in contributing to the limited development of a strong workforce. Therefore, in order to create a vital economic climate, the 2016 Statewide Town Hall suggested increasing the number of STEM degree graduates and transforming schools to a career guided system in which students can be encouraged to pursue a college degree and then a career. Additionally, the same document recommended implementing educational and training programs designed to enhance skills so as to transform the workforce. The TECH Center seeks to tackle the concerns of the 2016 Statewide Town Hall in New Mexico by stimulating the economy through the development of high-tech skills required for the STEM workforce.

NMC/CoSN and the TECH Center

The TECH Center addresses a number of points raised by the NMC/CoSN Horizon Report: 2017 K–12 Edition (Freeman et al., 2017). Freeman et al. (2017) stated that "as pedagogies that favor student-centered learning approaches continue to take hold across the world, tools such as VR that enable more experiential learning opportunities are increasingly valued" (p. 46). Responding to this point, the TECH Center at NMSU implements an experiential learning approach that involved students solving a problem through an aviation simulation. Additionally, Freeman et al. asserted that "the same technology used to simulate virtual experiences in medical and military training for years is now of interest to schools because it can provide students with simulated firsthand experiences" (p. 46). This statement deeply resonates with the purpose of the TECH Center, especially, since the setup of the TECH Center (see Figure 8.1) is funded and inspired by the Department of Defense.

To continue highlighting the conceptual concordance, Freeman et al. (2017) asserted that "a method of formalized problem-solving, computational thinking is a strategy that leverages the power of computers through collecting data, breaking it into smaller parts, and recognizing patterns" (p. 25). Students in the TECH Center engage in computational thinking by gathering data and entering information into the computer that will allow them to complete a specific task successfully. As highlighted above, for example, in the context of the aviation simulator, students are challenged to calculate how much fuel is required to travel from one location to another without crashing their plane, and program their flight to make a safe landing. Since computational thinking will soon be impacting a wide range of careers (Wing, 2008) from software developers to marketing, the

Figure 8.1.   The TECH Center at New Mexico State University incorporates programs that focus on critical thinking and problem solving in real world, challenges and scenarios.

Freeman et al. acknowledgment of the need to increase the development of computational thinking skills in students—which is one of the tasks addressed by the TECH Center—is timely.

TECH Center and the GISD Community

One of the many aspects that makes the partnership that is the focus of this chapter unique is the fact that a university is providing a rural school district with digital technology experiences to improve the conditions of their community. The SOC recognized how GISD required the resources to enhance the quality of their students learning and set about offering unique experiences otherwise not accessible to them. The partnership attends to the needs of a vulnerable population. GISD students and their families need an opportunity to hope that they can achieve more than what they currently have. Through the partnership, SOC endeavors to help break a cycle of oppression by offering students the opportunity to learn to their greatest potential.

There are many ways in which the TECH Center potentially impact the GISD community. First, GISD students are coming to the university to participate in the TECH Center. So far, TECH Center laboratories have been installed in four additional locations. The TECH Center at NMSU was the first lab installed in a university setting. The SOC at NMSU was carefully selected by the Department of Defense because of our partnerships with the school communities in New Mexico. Colleges can be a very scary place if one has no experience of them. By hosting them in an exciting and safe environment at the university, the SOC students are given the opportunity to see that the university is a friendly place and they become comfortable in the university environment. In the process of learning in the TECH Center, pertinent information about college can be provided to both students and their parents.

Second, the multigenerational cycle of poverty can be disrupted through the TECH Center, as students will become better equipped to join the workforce through their acquisition of high-tech skills. Lastly, students are being engaged in an authentic learning experience that will hopefully increase their interest in STEM and motivate them to pursue a career in a STEM-related field.

Casto (2016) declared that "schools can act as community centers by opening the school building to the community for use during non-school hours" (p. 142). In the case of the TECH Center, the university is opening its doors to GISD students. This partnering evokes similarities to other school-community partnerships in rural settings, which have been described by Casto as having enhanced "support for families, community development, and the development of a sense of place for children" (p. 157). By exposing students to a college-like experience, we are generating "social capital." Through partnerships we are providing an equal opportunity for students to gain exposure to a college environment in the anticipation that at least some will aspire to extend their success in school to the college level. Involving parents in their children's learning experience during OST can greatly motivate students to succeed in school (Ferrara, 2015). The participants in the SOC programs find voice within the collaboration. On the one hand, students are provided the opportunity to learn about STEM, teachers get to improve their pedagogical practices, and the parents gain the information and receive the support they need to motivate their children to improve their school performance. On the other hand, the university benefits by contributing to the community, and remains faithful to its fine reputation as a land grant institution.

The SOC recognizes that the economic well-being of the state is influenced by the preparation of the next generation. Therefore, the hopes and aspirations as a result of the collaborative endeavor with GISD range from increasing STEM interests in our participants to increasing academic achievement. As previously mentioned, the state of New Mexico ranks at the bottom in education quality. The SOC continues to work to redress this situation through its OST programs. Receiving support from the Department of Defense allowed the construction of the TECH Center. Through involving GISD students with the TECH Center, the SOC leverages the opportunity of grant funding from the federal government—specifically the DoD. We anticipate increasing our collaboration across schools within the GISD. Children in GISD deserve the opportunity to know that they can pursue a college degree and a STEM career. Most of the GISD students are not aware of the great opportunities available to them. The SOC provides professional development for GISD teachers in how to facilitate and encourage STEM learning and the development of each student's STEM identity. The SOC seeks to continue to foster the prosperity of the GISD community.

Lessening the Digital Divide and Enhancing the Opportunity to Learn

The TECH Center, as well as the other OST programs, provides students with an enriched high-tech academic experience that will hopefully impact their career choice. Ritzhaupt, Liu, Dawson, and Barron (2013) stated that "if individuals do not have access (to digital tools), they have less opportunity to use these tools for their personal empowerment" (p. 300). Students from low-income families are not as likely as their peers from higher-income families to have access to digital tools, and thus, will be less-proficient users of technology. The TECH Center at NMSU challenges the "digital divide" (Ritzhaupt et al., 2013, p. 291) by allowing students in southern New Mexico to use state-of-the-art technology. In order to heal the digital divide, Ritzhaupt et al. (2013) suggested creating partnerships between schools and community organizations, engaging students in out-of-school programs, and training teachers through professional development to integrate technology within their curriculum—all of which are practices being undertaken at NMSU's SOC.

By providing GISD with OST programs geared towards digital technology, students and their parents are enabled to develop a mindset that students in the community can seek a postsecondary pathway whether they continue at the Doña Ana Community College or at NMSU. Since NMSU is a land grant university and has a commitment to the community, the NMSU SOC offers a solid program. Further, students with OST program experiences will be prepared to face the challenges encountered in higher education. Through their immersion in OST programs, students have acquired digital technological knowledge, along with other disciplinary and collaborative skills crucial for college success. High-quality OST programs increase students' participation and achievement in school, and decrease the likelihood that students will join gangs, become victims or perpetrators of violence, or become teen parents (Ordonez-Jasis & Jasis, 2004). There-fore, the community benefits from the existence of OST programs because they contribute to the creation of safe environments for the personal and professional development of students.

LESSONS LEARNED FROM GISD-SOC COLLABORATION AND MOVING FORWARD

Based on the partnership experiences thus far, we must emphasize the need to seek out opportunities to openly engage in conversation with the community. Hearkening back to the earlier example of the SOC director, she established first name-basis interactions with the superintendents and the staff members within the GISD schools. Additionally, she initiated con-versations with the parents of the students involved in the OST programs. This established a nurturing interaction with the community. Having the children's families involved through the family festivals played a crucial role in impacting the community. We highly recommend that similar out-reach centers in universities across the nation to seek and maintain contact with the communities that they are serving and to enshrine in their practice the reality that the individuals being served are people with great potential. In some cases, challenging deficit thinking (Valencia, 2010) may be the first step towards establishing a meaningful partnership with rural com-munities. Mutual respect and appreciation between the staff of the SOC and the GISD has played a crucial role in establishing our healthy partner-ship among team members who seek to help communities in need. The GISD collaboration would not have been possible without the help of the current SOC assistant director, Nicole Delgado, and the program special-ists. Having the students' and their communities' best interests in mind is the foundation of our successful partnership.

As to what is next for the university and school district collaboration, the SOC-GISD partnership will be ongoing. The SOC will keep offering OST experiences that will impact the students' lives. The SOC continues to report to the superintendent on the success of events during the OST programs. The lessons learned during the partnership inform a set of guidelines for how the SOC will partner with other school districts. The guidelines highlight that, first, one must know the school district in order to effectively attend to its needs. Second, the lines of communication must be kept among all parties; thus, the people involved in the partnership must be in constant contact. It is important that the university staff/faculty talk to administrators within the school district. For example, the SOC holds a monthly meeting with the GISD superintendents so both parties can be kept up-to-date with the logistics of the OST programs. Lastly, each university participant must always keep in mind that he/she is a visitor in the school district, and remain constantly aware that it is thanks to the school district that the collaborative opportunity exists.

CONCLUSION

This chapter has described the collaboration between the STEM Outreach Center at New Mexico State University and the Gadsden Independent School District (GISD). The Gadsden community has been served throughout the years and has been positively impacted through the STEM out-of-school-time programs. The TECH Center at New Mexico State University has positively impacted the Gadsden community by providing the opportunity for the GISD students to experience state-of-the-art aviation simulators and to increase their digital literacy. Such learning experiences will hopefully motivate students to pursue a career in the STEM fields. Time and further research will tell how the TECH Center impacts students in their ongoing academic choices and academic performance. As the STEM Outreach Center fulfills its long-term dream to continue serving underrepresented students with scarce resources, it also hopes the students that it serves are fulfilling dreams of their own.

REFERENCES

Alliance for Science & Technology Research in America. (2018). *New Mexico's 2018 STEM & innovation report card* [Data file]. Retrieved from https://www. usinnovation.org/state/pdf_cvd/ASTRA-STEM-on-the-Hill-NewMexico2018. pdf

Auger, A., Pierce, K., & Vandell, D. L. (2013). *Narrowing the achievement gap: Consistency and intensity of structured activities during elementary school.* Presented at the biennial meeting of the Society for Research in Child Development, Seattle, WA. Retrieved from http://expandinglearning.org/research/vandell/resources/SRCD_Presentation_Final.pdf

Ball, C., Huang, K., Rikard, R. V., & Cotten, S. R. (2017). The emotional costs of computers: An expectancy-value theory analysis of predominantly low-socioeconomic status minority students' STEM attitudes. *Information, Communication & Society, 22*(1), 105–128. https://doi.org/10.1080/1369118X.2017.1355403

Blanchard, M. R., LePrevost, C. E., Tolin A. D., & Gutierrez, K. S. (2016). Investigating technology-enhanced teacher professional development in rural, high-poverty middle schools. *Educational Researcher, 45*(3), 207–220. doi:10.3102/0013189X16644602

Bransford, J. D., Brown, A. L., Cocking, R. R., & National Academy of Sciences/National Research Council, W. E. (2000). *How people learn: Brain, mind, experience, and school.* (Expanded Edition). Washington, DC: Department of Education.

Bruce, B. C., & Casey, L. (2012). The practice of inquiry: A pedagogical "sweet spot" for digital literacy? *Computers in the Schools, 29*(1), 191–206. https://doi.org/1 0.1080/07380569.2012.657994

Casto, H. G. (2016). "Just One More Thing I Have to Do": School-community partnerships. *School Community Journal, 26*(1), 139–162.

Chavez, A. M. (2018). NMSU STEM Outreach Center makes positive economic impact on Doña Ana County. *New Mexico State University News Center.* Retrieved from https://newscenter.nmsu.edu/Articles/view/12973/nmsu-stem-outreach-center-makes-positive-economic-impact-on-dona-ana-county

Children, Youth Families Department. (2014). *Gadsden Independent Schools early childhood data profile* [Data file]. Retrieved from https://cyfd.org/docs/Gadsden_School_District_Profile_.pdf

Dalal, M., Archambault, L., & Shelton, C. (2017). Professional development for international teachers: Examining TPACK and technology integration decision making. *Journal of Research on Technology in Education, 49*(3–4), 117–133. doi:10.1080/15391523.2017.1314780

Darling-Hammond, L., Zielezinski, M. B., & Goldman, S. (2014). Using technology to support at-risk students' learning. *Alliance for Excellent Education.* Retrieved from https://all4ed.org/wp-content/uploads/2014/09/UsingTechnology.pdf

Education Week. (2018). New Mexico earns a D on state report card, ranks 50th in the nation. *Quality Counts, 37*(17). Retrieved from https://www.edweek.org/ew/collections/quality-counts-2018-state-grades/highlight-reports/2018/01/17/new-mexico.html

Fayer, S., Lacey, A., & Watson, A. (2017). *STEM Occupations: Past, Present, and Future.* Retrieved from https://www.bls.gov/spotlight/2017/science-technology-engineering-and-mathematics-stem-occupations-past-present-and-future/pdf/science-technology-engineering-and-mathematics-stem-occupations-past-present-and-future.pdf

Ferrara, M. M. (2015). Parental involvement facilitators: Unlocking social capital wealth. *School Community Journal, 25*(1), 29–51.

Freeman, A., Adams Becker, S., Cummins, M., Davis, A., & Hall Giesinger, C. (2017). *NMC/CoSN Horizon Report: 2017 K–12 Edition.* Austin, TX: The New Media Consortium.

Krishnamurthi, A., Ballard, M., & Noam, G. G. (2014). Examining the impact of afterschool STEM programs. *Noyce Foundation.* Retrieved from https://files.eric.ed.gov/fulltext/ED546628.pdf

Legislative Finance Committee Program Evaluation Unit. (2016). *Science, technology, engineering and math (STEM): Degree production and employment outcomes.* Retrieved from https://www.nmlegis.gov/Entity/LFC/Documents/Program_Evaluation_Reports/Science,%20Technology,%20Engineering%20and%20Math%20(STEM)%20Degree%20Production%20and%20Employment%20Outcomes.pdf

McDonnell, L. M. (1995). Opportunity to learn as a research concept and a policy instrument. *Educational Evaluation and Policy Analysis, 17*(3), 305–322. https://doi.org/10.3102/01623737017003305

McKnight, K., O'Malley, K., Ruzic, R., Horsley, M. K., Franey, J. J., & Basset, K. (2016). Teaching in a digital age: How educators use technology to improve student learning. *Journal of Research on Technology in Education, 48*(3), 194–211. doi:10.1080/15391523.2016.1175856

New Mexico Department of Workforce Solutions. (2016). *A Report highlighting New Mexico's current and future workforce.* Retrieved from https://www. dws.state.nm.us/Portals/0/DM/LMI/NM_2016_State_of_the_Workforce_ Report_033016.pdf

New Mexico Public Education Department. (2016). *Gadsden independent schools district report card 2015-2016* [Data file]. Retrieved from http://webed.ped. state.nm.us/sites/conference/2016%20District%20Report%20Cards/019_ GADSDEN_INDEPENDENT_SCHOOLS_DRC2016_.pdf

New Mexico Public Education Department. (2017). *Gadsden Independent School District: School district report   card 2016–2017.* New Mexico Public Education Department. Retrieved from http://webed.ped.state.nm.us/ sites/conference/2017%20District%20Report%20Cards/019_GADSDEN_ INDEPENDENT_SCHOOLS_DRC2017_.pdf

Noonan, R. (2017). *STEM Jobs: 2017 Update* (ESA Issue Brief # 02-17). Office of the Chief Economist, Economics and Statistics Administration, U.S. Department of Commerce (March 30, 2017). Retrieved from http://www.esa.gov/reports/ stem-jobs-2017-update

Nugent, G., Barker, B., Grandgenett, N., & Adamchuk, V. (2010). Impact of robotics and geospatial technology interventions on youth STEM learning and attitudes. *Journal of Research on Technology in Education, 42*(4), 391–408. https:// doi.org/10.1080/15391523.2010.10782557

Ordonez-Jasis, R., & Jasis, P. (2004). Rising with De Colores: Tapping into the resources of la comunidad to assist under-performing Chicano-Latino students. *Journal of Latinos And Education, 3*(1), 53–64.

Ritzhaupt, A. D., Liu, F., Dawson, K., & Barron, A. E. (2013). Differences in student information and communication technology literacy based on socio-economic status, ethnicity, and gender: Evidence of a digital divide in Florida schools. *Journal of Research on Technology in Education, 45*(4), 291–307. https://doi.org/ 10.1080/15391523.2013.10782607

2016 Statewide Town Hall. (2016). *Economic security and vitality for New Mexico.* Retrieved from http://nmfirst.org/library/2016/2016-economic-vitality/index. html

U.S. Department of Education (2015). *Science, technology, engineering and math: Education   for global leadership.* Retrieved from https://www.ed.gov/stem

Valencia, R. R. (2010). *Dismantling contemporary deficit thinking: Educational thought and practice.* New York, NY: Routledge

Wing, J. M. (2008). Computational thinking and thinking about computing. *Philosophical transactions of the Royal Society A: Mathematical, physical and engineering sciences, 366*(1881), 3717–3725. doi:10.1098/rsta.2008.0118

Xue, Y., & Larson, R. (2015, May). STEM crisis or STEM surplus? Yes and yes. *Monthly Labor Review, U.S. Department of Labor, Bureau of Labor Statistics.* Retrieved from https://doi.org/10.21916/mlr.2015.14

Yarbro, J., McKnight, K., Elliot, S., Kurz, A., & Wardlow, L. (2016). Digital instructional strategies and their role in classroom learning. *Journal of Research on*

Technology in Education, 48(4), 274–289. https://doi.org/10.1080/15391523.2 016.1212632

Zheng, X., Stapleton, L. M., Henneberger, A. K., Woolley, M. E., & Maryland Longitudinal Data System Center. (2016). *Assessing the workforce outcomes of Maryland science, technology, engineering, and math (STEM) postsecondary graduates*. Retrieved from https://mldscenter.maryland.gov/egov/Publications/ STEMReport_Merged_8_2016.pdf

CHAPTER 9

COLLABORATING WITH EDUCATORS

Video Games Support Alternative Classroom Pedagogies to Support Boys' Meaning-Making

Carol-Ann Lane
University of Western Ontario

Scholars have acknowledged the contribution of video gaming to complex forms of learning, identifying links between gaming and engagement, experiential learning spaces, problem-solving, strategies, transliteracy reflectivity, critical literacy, and metacognitive thinking. Despite this movement toward the inclusion of video gaming in literacy teaching, concerns about certain risks raised by scholars have slowed the adoption of video games to foster learning. Existing research in video gaming practices, especially for boys, tend to focus on risks associated with boys' gaming choices. For example, boys who interact with video games and apply that knowledge to in-school practice are considered problematic due to scholars' significant reservations about stereotypical themes (such as themes of power, violence and misogyny) embedded in video game plots and characters. Despite this, research continues to emerge offering insights to educators that gaming and literacy are

Integrating Digital Technology in Education:
School–University–Community Collaboration, pp. 199–224
Copyright © 2019 by Information Age Publishing

not on opposite ends of the literacy learning spectrum but rather represent a highly unified multimodal foundation (Beavis, 2012; Gee, 2014; Squire, 2013; Steinkuehler, Squire, & Barab, 2012). Therefore, in this chapter I present some of my findings from my multi-case ethnographic study which examined the unique experiences of four boys engaged with video gaming in two different contexts: a community center and an after-school video club. This chapter also addresses how researchers can collaborate with school educators to support practical classroom strategies using multiliteracies resources, such as the Learning by Design framework (Cope & Kalantzis, 2016).

Using a multiliteracies lens, my multi-case study examined the experiences of four boys engaged with video gaming in two different contexts: a community center and an after-school video club. One of my main concerns was to protect the identities, backgrounds, and names of the participants in my study. In order to do this, I assigned pseudonyms to each of the boys. The pseudonym names of the four boys who participated in this study were Albert, Jeffrey, Mike, and Brian.

From my research, I have gained an understanding of the nature of boys' behavior and learning in social settings while they engage in video game play. Studying the ways in which boys make meanings through multimodal ways of learning can offer insights into strategies that can potentially reinvent traditional literacy pedagogical boundaries and establish new practices for building knowledge.

These ethnographic cases, along with their naturalistic aspects, strengthened the authenticity of the social-contextual-cultural experiences of the four, adolescent-aged boys and allowed me to gain an understanding of their everyday experiences. Interpretations of the cultural meanings made by each of the boys, based on their individual unique experiences engaging with video games, can provide readers with insights into how to approach adolescent aged boys' literacy development. The findings that emerged from my study indicated how these four boys developed their multimodal ways of learning by engaging with visual perspectives of video games. My methodological approach documented what the boys were saying, as much as possible, which is an aspect of boys' video gaming practices that is currently understudied in the literature. There were a number of findings emanating from this study, including the following: (a) boys use their video gaming practices for meaning-making and collaborative efforts in order to gain an understanding of several knowledge processes (such as decision making, predicting, analyzing, strategizing, etc.), (b) boys extend and apply their cultural knowledge as creative innovators, producing and publishing YouTube instructional videos for video game players and designing video games for a history project, (c) boys demonstrate peer mentoring through storytelling and face-to-face interactions or in their online community of practice, (d) boys make meanings using metacognitive literacy skills in a variety of ways, and (e) boys focus on cultural preservation and narrative storytelling. While acknowledging concerns related to video gaming, such as negative identity construction, violence, distraction, and time commitment for integration, this chapter seeks to contribute to the scholarly discussion about the use of video games in classrooms by explicitly considering the ways in which gaming may support boys' meaning-making and acquisition of cultural knowledge.

Some of the themes addressed by Freeman, Adams Becker, Cummins, Davis, and Hall Giesinger (2017) in the *NMC/CoSN Horizon Report K–12 Edition* (Horizon Report) indicate a future focus over the next five years to position pedagogical strategies—including learning and visualization technologies—to foster creative inquiry. Video games are not specifically named in the report which leads me to believe there still remains a certain reluctance by educators to adopt these, perhaps due to scholars' reservations about stereotypical themes (such as themes of power, violence, and misogyny) that may be embedded in video game plots and characters.

Adding to the complexity surrounding video games as an alternative strategy for learning, scholars' reservations become a challenge for researchers who collaborate with school educators to integrate out-of-school and in-school video gaming practices into the learning process (Cope & Kalantzis, 2009). Video gaming presents a new set of challenges to educators and learners for what literacy and multiliteracies mean, given the perspective of traditional literacy pedagogies in schools. This chapter provides an overview of the use of video games in classrooms by considering the ways in which gaming may support boys' meaning-making. Increasingly, scholars are recognizing how students need literacy practices in order to be prepared as future world citizens (Rowsell & Walsh, 2011). Multiliteracies is one way to embrace these new forms of literacy. Rowsell and Walsh (2011) explained that by invoking multiliteracies as a concept, it involves multiple genres, and multiple subjectivities that shift with context, texts, and the identities of people. Emerging research has the potential to inform secondary school English teachers regarding how to engage and motivate their students in multiliteracies.

APPLYING A MULTILITERACIES PEDAGOGY FOR THE LEARNING BY DESIGN FRAMEWORK

In order for the multiliteracies concept to be applied successfully in the literacy domain, The New London Group (2000) emphasized the need for a pedagogy to supplement what teachers do by considering "how the human mind works in society and classrooms, as well as about the nature of teaching and learning" (p. 30). Specifically, The New London Group (1996) addressed the question of how learners engage in common practices by applying four methods of instructional strategy: situated practice, overt instruction, critical framing, and transformed practice. With instructional strategy in mind, Cope and Kalantzis (2009) reconfigured The New London Group (1996) list to include experiencing, conceptualizing, analyzing, and applying.

Experiencing

Experiencing arises from the view that learners' cognition is situated, contextual and cultural (Kalantzis & Cope, 2012). Learners immerse in meaningful practices within a community of other learners. Kalantzis and Cope (2012) also recognized the pedagogical weaving between school learning and practical out-of-school experiences that are based on individual interests (see also Kalantzis & Cope, 2012). An important point with which I also agree and which needs emphasizing, is how these experiences interconnect culturally. Learners experience what they know by being reflective and by bringing their own experiences, interests, perspectives, and ways of understanding the world. Cope and Kalantzis explained how learners experience the new by being "exposed to new information, experiences and texts" (p. 18). They viewed this experiencing process, resembling a form of collaboration (Vygotsky, 1935/1978), as involving learners' willingness to take risks in a new domain of action and meaning. Learners transform meaning, but they also trust in the guidance of others, such as peers and teachers (Cope & Kalantzis, 2009; The New London Group, 1996).

Conceptualizing

According to Cope and Kalantzis (2009), the purpose of instruction is to focus the learners on significant aspects of their cultural meaning-making experiences and facilitate thinking or building knowledge within the community of learners. They drew on the cognitive learning theories put forth by Vygotsky (1978/1935) about how learners conceptualize their meanings by building mental models and abstract theories, becoming active learners in the process of knowledge building.

Analyzing

Analyzing is the part of the process by which learners establish relations between cause and effect and explain textual patterns and connections. It also adds a dimension to the knowledge process by extending the need for learners to constructively evaluate their learning and others' perspectives. Cope and Kalantzis (2009) emphasized the concept of learners creatively and reflectively innovating, but added that learners develop reasoning capacity by interrogating the interests behind a meaning, an action, or their own thinking process.

Applying

Applying involves learners demonstrating their acquired knowledge and applying it to real world situations. It represents how learners develop innovative and creative ways to demonstrate their meaning-making and knowledge (Cope & Kalantzis, 2009). This step also emphasizes the importance of learners' diversity, interests, and experiences. Cope and Kalantzis (2009) reminded educators that learners transform their meaning-making into other contexts by reflecting upon their own goals and values as they apply and revise what they have learned. Learners use their cultural insight and experiences to continuously revise their learning.

BACKGROUND

The Four Boys in my Study

Four adolescent boys (14 to 15 years old) participated in my dual-site ethnographic study of two cases each involving two boys, one case situated at a community center, the other focusing on an after-school video club. The two boys who were included in the community center case were Albert and Jeffrey; the other two boys who were included in the after-school video club case were Mike and Brian.

Community center. Upon meeting Albert, I observed that his demeanor appeared to be very quiet and that he never seemed to talk to anyone. He was a 14-year-old adolescent, tall, and of a thin physical stature. I observed him to be very courteous and respectful to me and to other students. The other students, not part of the study, I observed as being loud and very boisterous; at times, I noticed them pushing each other and yelling. Albert, on the other hand, was generally focused on his video gameplay and tended not to mix with the other youth at the center. For Albert, the community center did not appear to be a formal place where he received much useful guided practice related to either experiencing or conceptualizing (Cope & Kalantzis, 2009; The New London Group, 1996) for his designing and redesigning knowledge processes. More specifically, I observed his meaning-making, such as problem-solving and analyzing, as evidenced by his independent play and use of surrounding online networks or communities of practice. Alexander (2009) and Squire (2013) indicated that online communities of practice (Wenger, 1998), characterized by numerous message boards, represent ways for learners to actively build their literacy skills, such as onscreen writing, listening, critically reflecting, and thoughtfully responding. Albert explained to me that he would often play certain video games to practice his problem-solving and analytical skills and then access surrounding online networks to interact with other gamers or peers by demonstrating what he had learned independently. I observed that he did not share these designing and redesigning knowledge processes with peers at the community center.

The second participant in the community center case was Jeffrey. When I first met Jeffrey, a 15-year-old adolescent, I noticed that he was slightly shorter, but larger in physical stature than Albert. I observed that Jeffrey, like Albert, did not seem to be socially engaged with the other individuals at the community center. He frequently conveyed his gestural representation through his facial expressions, appearing to be frowning or agitated when other noisy students were in the computer video games room. A gestural representation is a metalanguage related to the physical body including facial expressions, movements of hands, arms, legs, dance, fashion, ceremony, and ritual, but also includes a form of feelings or emotions (Cope & Kalantzis, 2009, The New London Group, 1996). I observed Jeffrey to be very quiet and introspective in his actions. Although observations of Jeffrey's personality would be difficult to justify and would require further investigation for a future study, I frequently noticed in later observations and casual conversations with him that he preferred not to interact much with others. The community center did not seem to offer Jeffrey opportunities for the ways he made meanings during his learning processes—either experiencing or conceptualising (Cope & Kalantzis, 2009; The New London Group, 1996). During observations and conversations, Jeffrey revealed aspects of his mastery of several video games, practices, and particular discourses such as storytelling—an example of oral language (Cope & Kalantzis, 2009).

Jeffrey was highly articulate in his conversations with me, often telling stories about the video games he played, thereby providing an example of drawing on oral language representation (Cope & Kalantzis, 2009). Oral language refers to the linguistic element of the available designs which include speaking, audio text, music, sounds, hearing and listening (Cope & Kalantzis, 2009). The community center setting became an important context for Jeffrey's behaviors. He always appeared to be conscious of his surroundings (Wolcott, 1987)—the room, its function, and people interacting around him. He would often evoke gestural representations with his abrupt body language: walking into the video games room, turning his body, and then leaving if other adolescents or children were in the room.

After-school video club. The third participant, Mike, was a tall, 15-year-old adolescent, with a thin physical stature, and was the president of the after-school video club. I observed him as friendly and respectful to me and to other students in the club. Mike chatted with everyone, ensuring all were enjoying playing the games. During my observations, Mike generally focused on his gameplay but still found time to mentor other players. He would frequently pause a video game sequence during a competition to instruct other players on gameplay functions, strategies, or problem-solving. Thus, the after-school video club appeared to be both a formal and an informal place for Mike providing him opportunities to experience and conceptualise with others (Cope & Kalantzis, 2009; The New London Group, 1996).

Mike's meaning-making also emerged from his rich collaborative interactions in the online surrounding networks and communities of practice (Alexander, 2009; Squire, 2013). He constantly exchanged strategies, ideas, and best practices with others who shared a connection and common video gaming goal with him (Gee, 2007). Being the president of the club, Mike took up his gestural representation (Cope & Kalantzis, 2009) by walking ahead of everyone, and setting up the room to organize the activities. He wheeled in the televisions and console/video cassette recorder, turned on the widescreen overhead projector and selected the teams for gameplays.

The fourth participant, Brian, was a 15-year-old adolescent, shorter and slightly larger in physical stature than Mike. I observed Brian to be friendly and respectful to the other students, but he did not appear as talkative as Mike. Brian frequently conveyed his gestural representation (Cope & Kalantzis, 2009) by quietly walking around the room, observing others, and asking Mike when he could be the next player in a particular gameplay sequence. During video gaming events, Brian was quick to share his meaning-making experiences with fellow peers by explaining certain gameplay sequences. Brian appeared to be conscious of the high-energy surroundings in the ways he demonstrated his gestural, visual, and spatial representations (Cope & Kalantzis, 2009; Wolcott, 1987). Cope and Kalantzis (2009) redefined the visual representation to include video gaming audio text and on-screen digital text, still or moving images, view, and scene or perspective; the spatial includes proximity, spacing, layout, interpersonal distance, territoriality, architecture and landscape. During his game playing experiences, he would draw on video game characters' ability to jump, dance, and perform acrobatic movements on the computer screen. He would sometimes physically perform these same movements in the classroom in front of his peers, and did not appear inhibited by the surroundings, his peers, or the space he occupied.

In some ways, Brian demonstrated weaving his experiences with peers (Cope & Kalantzis, 2009), by openly demonstrating unfamiliar actions and texts to others. Brian appeared to be an expert in video gaming, and he would often find ways to mentor other players and share experiences. He would pause a video game sequence and guide other players in the mechanics of characters and gameplay strategies.

RESEARCHER/COMMUNITY/SCHOOL COLLABORATION

As a university researcher, one of the factors guiding my decision to focus on community collaboration was the way that some schools can represent additional levels of complexity for students in the forms of authority, classroom rules, peer-based social dynamics, and prescribed curriculum learning outcomes (Connell, 1996; Lingard & Douglas, 1999; Mac an Ghaill, 1994).

Gaining Access

First, as a researcher, my endeavor to collaborate with the community and schools began with finding opportunities at Ontario youth community centers to recruit boys to participate in my study. I was invited by a director of a rural community center to give a presentation to boys at the youth center about the study I was conducting. Second, as part of my recruitment process, I also contacted several Ontario school board principals to find out if any of the schools were running after-school clubs relating to video games and if I would have an opportunity to recruit boys who were members of those clubs. In addition to building parental trust and gaining parental consent, the school board's approval was necessary for me to have access to both the after-school video club and the community center.

Further, gaining entry to the community center involved securing the approval of a number of gatekeepers or key informants, including the manager and the program administrator of the community center, and the youth activity coordinator. Gaining the approval of the gatekeepers added credibility and legitimacy to me as a researcher in conducting my fieldwork because they were trusted by parents and the gatekeepers' approval resulted in my gaining permission to visit the community center and after-school club on a daily basis (Patton, 2002).

ARE BOYS UNDERACHIEVING IN LITERACY?

The Ontario Ministry of Education (OME, 2004) reported that literacy rate for boys lagged the rate for girls. Other literacy agencies have similar findings, including Canadian, American, and other international agencies. To gain a deeper understanding of this achievement gap, I examined a number of recent literacy assessment reports. In a report issued for the Ontario Education Quality and Accountability Office (EQAO, an independent agency funded by the Provincial Government of Ontario), Brochu, Deussing, Houme, and Chuy (2013) identified trends indicating that "over the past 5 years fully participating females (students who actually wrote the test) successfully achieved the literacy test on average at 87.4%, as compared to fully participating males with a successful achievement level of 80.2%" (pp. 67–68). Previously, Brochu, Gluszynski, and Knighton (2010) reported that females outperformed males by 34 points on the Programfor International Student Assessment (PISA) test in 2009. Statistics Canada further analyzed the trend for Canadian PISA test results and indicated that across Canada, the gender gap showed no statistical changes from 2009 to 2015 in the magnitude of the reading gap favoring females (Organzation for Economic Co-Operation and Development [OECD], 2015, pp. 41 & 44). In addition, the 2012 PISA results (issued in December 2013) and the 2015 PISA results (issued in December 2016) both reported a similar trend of girls performing considerably better than boys in literacy skills in all OECD countries from the year 2000, and in all Canadian provinces (OECD, 2013, 2016). The U.S. literacy results for 2012, reported by the National Center for Education Statistics (NCES, 2013) on the National Assessment of Education Progress (NAEP) follow the same pattern, revealing a 40-year trend in a gender gap which has favored females over males.

Despite these global standardized tests results being reported at aggregate levels, North American policymakers have continued to rely upon them to understand literacy achievements. Despite the long term reported trends of boys' underachievement, scholars continue to question the reliability of literacy scores from standardized tests because these are reported at aggregate levels without consideration of social economic status, race, and geographic location (see for example Martino & Rezai-Rashti, 2013). Critiquing the distortions that they saw emerging from problematic comparisons in the use of the Programme for International Student Assessment (PISA), Martino and Rezai-Rashti (2013) raised concerns that "framing of boys' literacy underachievement data … lends itself easily to being co-opted by a recuperative masculinity politics and, in so doing, tends to eschew important questions of class privilege that exists [sic] for middle-class boys" (p. 600).

Although there is complexity surrounding the reliability of aggregate reporting, some positive steps are being taken by the NCES to differentiate which boys are underachieving on the NAEP assessments. In its review and analysis of boys' literacy underachievement, the NCES has attempted to disaggregate the standardized test results within the race category. Based on this, NCES (2013) found that over the 40-year trend for 13-year-old Caucasian boys showed no change in their achievement levels as compared to black and Hispanic male students, who improved in their literacy achievement. Furthermore, one of the ongoing concerns for the Ontario Ministry of Education (OME) is to promote action with respect to boys' underachievement in literacy with their regular promotion of boy-specific literacy programs. As part of this ongoing concern, the OME has made various resources available to teachers by making available in excess of twenty specific boys' literacy publications, teacher inquiry sites, digital videos, and work plan support booklets for teachers' practical use (Ontario Ministry of Education, 2004, 2009a, 2009b). However, some scholars argue that the featured literacy strategies for boys may represent a return to patriarchal approaches to schooling (Lingard, Martino, & Mills, 2009; Lingard, Martino, Mills, & Bahr, 2002; Martino, 2013). Several scholars have critiqued the gendered concerns related to achievement (Frank, Kehler, Lovell, & Davison, 2003; Greig, 2003; Kehler, 2007, 2010; Lingard, Martino. & Mills, 2009; Lingard, Martino, Mills & Bahr, 2002; Martino, 2013), often dismissing policy guides as part of the boy-friendly strategies.

I acknowledge the concerns raised by scholars, however my interpretation of North American and global literacy results indicates that boys' literacy achievement still remains a concern. More importantly, I am interested in the recommendation made by Brochu et al. (2013) that "it might be possible to harness boys' performance in digital reading to improve their reading proficiency in both print and digital formats" (p. 43). In response to the OME's (2009a) recommendation to focus on digital literacy, in this chapter I explore how some boys relied on visual spatial skills to demonstrate learning outcomes as they made meaning by interacting with video games. For these reasons, my research is situated around the importance of considering alternate pedagogical strategies, such as video gaming.

Multiliteracies Meaning-Making: Four Boys' Experiences Engaging With Video Games in Their Designing Processes

The findings emerging from my ethnographic study illustrated a number of ways that these four boys (Albert & Jeffrey, the community case; Mike & Brian, the after-school video club case) made meanings in their designing processes as they engaged with video games. These four boys' meaning-making relates to learning processes found within the multiliteracies framework (Cope & Kalantzis, 2009; The New London Group, 1996). The meaning-makers use designs or representations in their designing processes to shape, create, and innovate meanings, based on their contextual and cultural experiences.

Table 9.1 represents the various domains or themes that emerged from my study. It illustrates an initial organization of cultural terms identifying domains or themes within the cases. A sample of participants' exact words provides a description to each cultural domain.

Table 9.1.
Cultural Domain Taxonomy

Cultural Domains	Sample Cultural Terms Uttered by Participants
Decision-making	A lot of decision-making; thinking ahead too; different decisions to get different outcomes
Problem-solving	Teaches you how to analyze things; sort of helps them in solving problems; good at puzzles; build different things
Learning strategy	Being smart in, I prefer strategy; reading your opponents; tools, better tools, mechanics; predict psychological aspects
Training, teaching others, teamwork	Plot and characters I talk a lot to people about; where your peers are the best training partners; I mean collaboration; I try super hard to have them cooperate but it never works, with friends you know them, you can cooperate with them
Exploring or experiencing emotion by interacting with narrative-focused games	This game is interesting because it explored this feeling; I really enjoy games like that explore metagame ideas, or like emotional ideas—interactivity; the game sort of mixes emotional storytelling along with great gameplay; emotions pushing the boundaries of games; change story, nonlinear storyline
Sharing moments by learning and understanding	Plot and characters, I talk a lot to people about; I'm gonna be the one to creating, game related things, it's usually like thinking about mechanics; I'll share them with someone who is capable about stories; helps me to learn and get better; took this cliché from this other game, just turned it on its head, made it five times more interesting; you learn about that which could help with literacy
Choice of characters – gender, violence	Choosing characters ahhh has never been ahhh a question of masculinity; I would either choose like either gender; I don't like violent games … umm yeah I'm actually proud that I don't like that stuff; I don't consider myself a particularly masculine person anyway
Cultural insights – accepting community, online community of practice	Parts of the game—cultural insights, little videos you would watch…. Elders come up to a camera; they would talk about the wisdom, the Elders; I actually met a lot of friends through Melee; pretty much my best friends are all in the video game club

Cognitive Meaning Making: Decision-Making, Problem-Solving, and Learning Strategy

As a self-initiated learner, Albert (the community center case) would play puzzle platformer video games and then transition to playing detailed strategy multiplayer competitive games, interacting with online players to experience the new and develop his problem-solving, strategy, and decision-making skills. Thus, Albert focused on different ways of thinking about literacy, such as spatial skills, based on "divided visual attention" (Gros, 2007, p. 29), by developing strategies about decision-making and problem-solving.

Albert's actions parallel suggestions made by some scholars that meaningful learning, including critical thinking, problem-solving and decision-making, can occur during gameplay sequences (Hommel, 2010). He relied on certain gameplay sequences and game type preferences to develop his problem-solving, strategy, and decision-making skills, perhaps through trial and error, in an experiential learning space (Squire, 2013). For example, on one occasion I found Albert playing his usual start-up video game, Portal 2 (Valve Corporation, 2011). While he was playing this game, he was eager to confide that he had trouble problem-solving, but that he had figured out the puzzle by continually playing this video game. Albert provided an example of this theory by indicating, "That's how you do it. I was having trouble figuring that out on my own even over that last few days. I've been doing this wrong. I figured that on my own." So rather than ask for help from other players at the community center or consult online help boards to draw on the available community of practice, Albert alternated between the conceptual methods of problem-solving and experiential methods of trial and error (Cope & Kalantzis, 2009).

By making use of experiential learning spaces present in video games, Albert could attempt several different methods and make meanings until he was satisfied with his learning goals (Apperley & Beavis, 2011; Gee, 2003; Squire, 2013). He would carefully analyze every step he took in the video game to solve a problem or redesign a new strategy. Albert played puzzle platformer games by applying and weaving different knowledge processes, and from experiencing these gameplay sequences, he developed skills to be self-reliant in problem-solving (Cope & Kalantzis, 2009).

Jeffrey made meanings by shifting between available modes in the video game, Never Alone (Kisima Innitchuna) (Upper One Games, 2014). Jeffrey experienced Alaskan heritage, as expressed by Elders through interactive videos embedded within the video game. For example, he viewed the Elder videos between the gameplay sequences, representing a good example of synaesthesia. This term is key to understanding modes of representations. Even if meaning-making is focused on one mode, it is still intrinsically multimodal with sound, images, and text side by side (Cope & Kalantzis, 2009). Jeffrey's interaction with these elements represents how "conscious mode switching makes for more powerful learning" (Cope & Kalantzis, 2009, p 181).

Jeffrey's experiences playing these types of narrative-based games also suggest that he is a reflective thinker, weaving and developing his cultural knowledge along with understanding other diverse cultural identities (Ajayi, 2011; Apperley & Beavis, 2013; Beavis, 2012; Cope & Kalantzis, 2009; Newkirk, 2002). Jeffrey's reflectivity is an example of Squire's (2013) argument that video game design "allows learners to share stories, theories, and experiences with their products, further tying the learning experience to their work outside the learning context" (p. 115). He also demonstrated his literacy skills by making meanings about the contextual significance of the storyline, which Duncum (2004) and Jenkins (2002) argued is a process by which learners decode and interact with the text and then link background experiences to new experiences to gain knowledge. Jeffrey demonstrated his cognitive awareness of the interplay of events and characters in the way he interrogated the author or developer's perspective. This is a good example of the explanation offered by both Beavis (2012) and Jenkins (2006) of how meaning makers must have a familiarity with the back-story of a video game to contribute to their knowledge processes. Jeffrey also demonstrated his literacy skills of making inferences and drawing conclusions about characters and his keenness to follow the storyline (Cope & Kalantzis, 2009; Pillay, 2002).

Informal Techniques of Social Dynamics—Conceptualizing

One of the distinguishing factors shared among the boys was the way in which they demonstrated active concept and theory making (Cope & Kalantzis, 2009) by adopting and sharing knowledge processes through social interactions during their cultural experiences. These collaborative actions, words, and reactions in their settings were classified in the training partners domain. Mike and Brian conceptualized their thinking processes for solving problems and developed strategies to co-produce knowledge through exchanging ideas with other members of a team or partners in a tournament. Their actions mirrored Gee's (2007) description of players engaging in gameplay on level playing fields through knowledge sharing and collaborating. What differentiated Mike's approach to solving problems was an emphasis on quick, correct decisions. Mike, and indeed Brian, collaborated with others, at least when they played the multiplayer competitive strategy game called Super Smash Bros. Melee (HAL Laboratory, 2014). They played with partners in order to read their opponents psychologically and gain an understanding of how to "beat them." Their learning processes resembled Vygotsky's (1935/1978) zone of proximal development in the way they observed and collaborated with more capable peers.

Social dynamics were demonstrated in my study, occurring in different forms with Jeffrey, Mike, and Brian. Jeffrey played video games online using surrounding networks of random players. For Jeffrey, playing with friends was an important aspect of his cultural gaming experiences because he knew their play style and had fun sharing ideas and questions with them. Well-known learning theorists including Piaget (1972/2008) and Vygotsky (1935/1978) have suggested that children develop certain behaviors to guide themselves. They organize their own activities, within a favorable environment, to fit a social form of behavior oriented to success.

Brian's approach to cultural knowledge was similar to Jeffrey's in the sense that he preferred sharing exciting moments with others during these social interactions. Brian called these moments positive fleeting interactions, found in the learning— sharing moments domain. Brian explained his experiences when he was gaming with peers to be "like fleeting positive interactions with people ... everyone gets excited about it that second and then like the game continues." He also explained that peers offer support and are the best training partners. Brian's demonstration of video gaming as having a social impact on his learning mirrors Steinkuehler, Squire, and Barab's (2012) findings—in accord with Vygotsky's (1978/1935) insights into learning as socialization—when they argued "consistent with the sociocultural approach, it's equally important for researchers and theorists to understand the socially situated nature of gameplay" (p. 10). This view coincides with Gee's (2007) perspective of building "affinity spaces" (p. 91). Although affinity spaces relate more specifically to online gaming, Gee's views also apply to video gaming in the same way, as evidenced by Brian's perspective about playing with a human, which provides more depth to learning rather than just playing against a computer. Gee argued that

> players can play alone against the computer or with and against other human players. Whether they play alone or together, the enterprise is social since almost all players need to get and share information about the games in order to become adept at playing them. (pp. 91–92)

The actions and responses demonstrated by Mike and Brian revealed an emerging theme of peer mentoring and collaboration. Although Brian and Mike were stereotypical gamers, showing much enthusiasm, and even appearing, at times, to be in competition with each other during their tournaments, they always found the time to provide guidance, and peer-to-peer teaching to other players. Peer mentoring supports Steinkuehler, Squire and Barab's (2012) claim that, "people get encouragement from an audience and feedback from peers, although everyone plays both roles at different times" (p. 144).

Moreover, Mike and Brian demonstrated this preference for bonding and socializing during their video game playing sequences by physically sitting close together, cheering, dancing, clapping their hands, and encouraging each other. When Albert and Jeffrey recounted their video gaming experiences with friends, they also shared how they built relationships, helping the team and each other. These forms of social interaction support Vygotsky's (1978/1935) theory that learning is an external process.

EXPLORING OR EXPERIENCING EMOTION

Many times, Jeffrey referred to a video game called The Last of Us (Naughty Dog, 2013) to convey his cultural terms: "It's in the game ... sort of mixes emotional storytelling along with great gameplay." Jeffrey's reflection referred to the emotion built into the video game but did not diverge too much from his common or preferred experience of storytelling. He also made further connections to emotion-based stories in his explanation of his experience playing Valiant Hearts: The Great War (Ubisoft Montpellier, 2014), another historically based adventure video game. According to Jeffrey, "there's quite a bit of emotion in that game that was probably the last time I've ever had ... like emotion for a video game." He also focused on having dialogues with friends and contributed meaningful questions about the experiences he encountered when he played these particular video games.

Brian explored metagame ideas (the degree to which a video game's characters and plot can evoke emotion from a player) by playing games with in-depth storylines and emotional depth of characters. Brian explained, "You get to like really live with the characters like Undertale [Fox, 2015, a role-playing game] would be like an adventure game— you experience things as things happen." Brian preferred to engage with the immersive quality of games to provide him with emotionally compelling experiences. He explained, "Soo primarily I'll play very story driven Indie games or ummm, very ahhh very narrative focused or centred around parts of a game's sort of emotions that trivial games don't normally explore." Brian was an active theory maker, building his knowledge processes by interpreting and exploring these experiential types of adventure games.

Cultural Insights

Jeffrey focused on emotional interactive video games, such as Never Alone (Kisima Innitchuna) (Upper One Games, 2014), where he experienced Alaskan heritage, as expressed by Elders through interactive videos embedded within the video game. He also found that cultural meaning emerged for him in an adventure video game called Valiant Hearts: The Great War (Ubisoft Montpellier, 2014). In playing this game, Jeffrey was able to engage with the characters (VanSledright, 2002) and a nonlinear storyline to understand history and experience emotional reflectivity (Alexander, 2009). Jeffrey developed cultural meanings from playing video games in which he could change plot sequences or which had a nonlinear storyline (Jenkins, 2006; Squire, 2013). His video gaming experiences were more meaningful to him than reading a linear story in a book or playing a puzzle platformer game, which lacked a storyline.

Jeffrey showed a strong preference for playing Never Alone (Kisima Innitchuna) (Upper One Games, 2014). He appeared to focus on his oral skills as a storyteller (Steinkuehler, 2007). In playing this game, he experienced the known from viewing embedded authentic videos created by Alaskan Elders providing their wisdom through storytelling in real world situations (Cope & Kalantzis, 2009). Jeffrey experienced the new through his awareness of the Alaskan culture, and he demonstrated this awareness by suggesting the need to preserve it and observing the lack of cultural and diversity competencies in schools (Ajayi, 2011). In a sense, his association with the Alaskan culture within the video game parallels findings from Jenkins (2002) who argued that games designed with embedded literary genres can induce players to make pre-existing narrative associations from games enacting certain narrative events. Jeffrey made use of the dynamic interplay of visual and audio representations (Cope & Kalantzis, 2009) by actively engaging with the video game-embedded authentic video elements. He explained,

> Parts of the game Never Alone ... Cultural insights ... these little videos you would watch ... Elders come up to a camera. They would be interviewed by the developers ... They would talk about like their stories ... They would talk about the wisdom, the Elders.

Jeffrey demonstrated his own perspectives on culture in his meaning-making comments: "Yeah ... I know I just love the basic idea going from a native's perspective." Many of his comments were related to the Never Alone (Kisima Innitchuna) (Upper One Games, 2014) video game. Jeffrey applied his creativity and perception, cultural knowledge, and understanding of the complex diversity from playing this game to a real-world situation by suggesting integrating the Ojibway language in school curriculum. He did not self-identify as Ojibway, but he appeared to have extensive knowledge of this culture by making frequent reference to its relevance. In addition,

Jeffrey appeared to be knowledgeable of other cultures, specifically First Nation communities, and the fact that these cultures warrant preservation.

Gender and Violence

Toxic masculine behaviors were absent from my results as those behaviors were never exhibited by the boys. Mike and Brian preferred to adopt animal characters, non-masculine characters, or feminine characters when they played video games. They explained how they made these choices on a regular basis and were quite open with peers in tournaments and/or public competitions. They provided context in their post-observation interviews by citing reasons for choosing feminine characters as stronger than masculine ones, and generally choosing nonviolent options in video games. They also explained that their game choices were oriented to not harming other characters during gameplay sequences. The only evidence emerging from this theme about gender was with Albert, who admitted to exploring feminine and masculine characters. He explained that he was interested in understanding the abilities of those characters so that his understanding would help him to play the game differently or respond differently to the gameplay experiences. Albert admitted to selecting his gender, a male character, first then during his game play sequences he would select a different gender, a female character. Albert's response challenges Sanford and Madill's (2006) claim that boys tend to "resist traditional school literacies, choosing instead modes of literacy to support the particular type of masculine persona they have selected for themselves, and make a commitment to that self-selected identity" (p. 299).

A significant emergent pattern was how the boys openly rejected violence or violent video games. By playing games targeted for younger players, such as Nintendo's Super Smash Bros. Melee (HAL Laboratory, 2014), both Mike and Brian made conscious efforts to reject violent video games. Mike provided evidence by specifically admitting to having a phobia about blood and not liking violent video games. Furthermore, Mike and Brian both provided context about not playing violent video games and would not maintain friendships with peers who preferred these types of games.

Similarly, Jeffrey openly rejected playing some video games at the center because they contained violent themes. His resistance was further justified by his criticism of other players who played these violent themed video games. He preferred playing story-based video games that provided him with knowledge about history or Indigenous cultures. For example, he was entirely opposed to using any violent video games, and continuously asked to play the Aboriginal-based video game Never Alone (Kisima Innitchuna) (Upper One Games, 2014). His actions challenge the notions put forth by Sanford and Madill (2006), who claimed that video games provide players with opportunities "to demonstrate their heterosexual masculinity" (p. 297).

Outward rejection of violent video games occurred in three of the four participants in my study, but Albert showed a greater preference for point/ shoot video games. However, my findings from Jeffrey, Mike, and Brian challenge Mac an Ghaill's (1994) notion that video-gamers belong to a student micro culture which amplifies a masculine stereotype through the members forms of resistance. Clearly, three of the four boys resisted stereotypical behavior by not fitting into the forms of resistance, as Sanford and Madill (2006) have suggested.

MEANING-MAKING FROM VIDEO GAMING PRACTICES

Albert, Jeffrey, Mike, and Brian demonstrated a number of meaning-making processes as they engaged with their out-of-school video gaming practices. This suggests several implications for educators who may want to integrate alternative resources in the classroom to leverage practices from nonacademic learning domains such as video gaming in addressing boys' underachievement in literacy. Similar to recent studies conducted on multiliteracies—including video gaming practices (see for example, Apperley & Beavis, 2011; Ganapathy, 2014; Gee, 2003, 2007; Sanford & Madill, 2007; Squire, 2013; Steinkuehler, 2007, 2011; VanSledright, 2002)—my study suggests ways for teachers to establish new practices for building knowledge by engaging 21st century learners in multimodal meaning-making and knowledge processes, particularly through the use of video games. The findings from my study may also inform classroom instruction by supporting professional teacher development if teachers choose to develop games aligned to their curriculum objectives.

Making Meaning

Mike and Brian made meaning by using their metacognitive skills. Conceptually, they named their knowledge processes by identifying and engaging in logical steps to achieve their goals. First, they would find opportunities to play video games with human partners, determined by Brian to be optimal training. Second, they appeared to follow a routine logical process to conceptualize and put their theories into practice. By relying on metacognition for meaning-making, they developed their understanding and perception of gameplay strategies by reading opponents. Their meaning-making involved observing small details, understanding the characters and game functions, identifying psychological aspects of their opponents, and building on these concepts to develop their own strategies.

Albert developed his conceptual theories by independently playing video games, developing strategies, and building his skills. Once he felt he had mastered his skills, using certain or building strategies through games like Minecraft (Mojang, 2011), he extended his theories to his online community of practice by showing them how to play (Gros, 2007). These activities highlight Albert as an active theory-maker (Cope & Kalantzis, 2009) by extending his learning processes with peers (Jansz, 2005). Using video games such as Minecraft may seem like a diversion from learning literacy, but Kafai and Burke (2016) advocated these types of games because they promoted meaning-making and invited players to develop skills in designing and creating.

Some of the ways that Jeffrey demonstrated analysing and reading for meaning was through playing nonlinear, interactive, narrative-type video games. Jeffrey analyzed narrative video games functionally by identifying the differences between video games and novels, explaining that novels provide the same storyline, unchanged, whereas interactive video games provide more engagement with the storyline because players can alter it based on character and plot choices. Jeffrey demonstrated his understanding and analysis of narrative genres by immersing himself into the game experience and co-constructing the narratives (Jenkins, 2002; Squire, 2013)—he gained literacy skills by demonstrating his use of alternative texts in nonlinear, interactive video games (Jenkins, 2002; Sanford & Madill, 2007). These nonlinear video games also invite learners to be creative in their meaning-making by altering storylines based on their values and perspectives.

Analytical Skills

Jeffrey demonstrated his critical analytical skills by reflecting on small details about video game plot and characters. In playing The Last of Us (Naughty Dog, 2013), he identified a character named Bill and questioned his convenient placement within the plot to solve problems. Jeffrey organized his understanding about storyline plot and characters by analyzing meanings and critically framing his understanding of the plot and character connections (Cope & Kalantzis, 2009). Clearly, Jeffrey demonstrated his skills in critically evaluating relationships and the development of characters in a storyline. In this way, The Last of Us offered Jeffrey opportunities to develop his critical perspectives about how texts work, helping him to identify assumptions built into texts—such as characters' roles within the storyline—and how this related to his own perspectives and culture (Beavis, 2012; Cope & Kalantzis, 2009). These activities may provide opportunities for students to become more engaged in a variety of texts and help to anchor their out-of-school interests.

Making Connections

Brian explored metagame ideas such as emotion. As pointed out previously, Brian's video game preferences focused on experimental games, such as Undertale (Fox, 2015), and multiplayer competitive games such as Super Smash Bros. Melee (HAL Laboratory, 2014), similar to Mike's preferences. Brian's choice of immersive video games enabled him to creatively make meaningful connections and explore emotions (Cope & Kalantzis, 2009; Squire, 2013). He preferred games that pushed the emotional boundaries, thus demonstrating his metacognitive ability to evaluate, interpret, and express his ideas to others. Brian demonstrated his skills and understanding of emotional boundaries in the ways he thought about characters through playing an interactive, narrative video game. Introducing interactive, narrative video games in the classroom, such as Undertale, may enable students to experience the same type of metacognition as experienced by Brian.

CAN VIDEO GAMES ADDRESS BOYS' LITERACY UNDERACHIEVEMENT?

One of the important aspects of deep learning is how digital tools, such as video games, can support connections students make based on their interests and daily lives. In its definition of educational technology, the Horizon Report (Freeman et al., 2017) focused on digital strategies and how these could be used to foster meaningful learning. According to the Horizon Report, some important developments for digital strategies include

> ways of using devices and software to enrich teaching and learning, whether inside or outside the classroom. Effective digital strategies can be used in both formal and informal learning; what makes them interesting is that they transcend conventional ideas to create something that feels new, meaningful, and 21st century. (p. 38)

Moreover, the innovative section of the report identifies digital strategies as games and gamification. The Horizon Report (Freeman et al., 2017) did not specifically give attention to video games as an alternative pathway to learning, however it did suggest that the use of digital strategies for both formal and informal learning was essential to meet the needs of 21st century learners. The Horizon Report also indicated how educators could introduce strategies for learners to be creative in their inquiry based learning practices and how technology could support those strategies in calling on "educators around the world to focus on their teaching strategies such that students lead their own inquiry" (p. 14).

Moreover, the Horizon Report (Freeman et al., 2017) suggested that "teachers must also acknowledge the prior experiences students bring to the classroom, helping them integrate and transfer knowledge to new situations, and support their ability to be aware of their learning and gain confidence in their solutions" (p. 14), and provided the following sobering reflection:

> As schools prioritize active learning over rote learning, students are being viewed in a new light. The embedding of maker culture in K–12 education has made students active contributors to the knowledge ecosystem rather than merely participants and consumers of knowledge. They learn by experiencing, doing, and creating, demonstrating newly acquired skills in more concrete and creative ways. Students do not have to wait until graduation to change the world. However, schools continue to be challenged to generate these opportunities in spaces and with paradigms that still lean on traditional practices. (p. 8)

One of the ways learning can offer insights into the reinvention of traditional literacy pedagogical boundaries and establish new practices for building knowledge is through multimodal meaning-making. Part of the shift to prioritize student-centered learning practices whereby students are active contributors to their creative knowledge processes can be found in video game practices. Steinkuehler, Squire, and Barab (2012) suggested how learning evolves from the interaction between the game play and the social practice that occurs in and around video game practices. They further explored the social function of video gaming practices by suggesting that a broader framework surrounds video games and helps to facilitate individual and collaborative game play and thus learning. Despite the advantages of including video gaming in literacy teaching, however, scholars have raised concerns regarding the imperative to simultaneously build awareness of the school community (see for example, Ajayi, 2010; Alexander, 2009; Sanford & Madill, 2007; Skerrett, 2011).

RECOMMENDATIONS AND
THE LEARNING BY DESIGN FRAMEWORK

The Learning by Design framework (Cope & Kalantzis, 2016) invokes multiple modalities of meanings and a range of knowledge processes (experiencing, conceptualizing, analyzing, and applying) so that practitioners will be enabled to prepare lessons with multiple curricula objectives and differentiated activities for a broad range of learners. It is a flexible framework, allowing learners to draw from any of the available design modes, such as oral, visual, audio, tactile, gestural, and spatial, to make meaning (Cope & Kalantzis, 2009). This framework does not prescribe a correct pedagogy, but rather provides the language with which to interpret and define features around any pedagogy that a practitioner employs. Specifically, the Learning by Design framework consists of concept labels applied to the different stages of a learner's experience (Cope & Kalantzis, 2016). These labels are defined as knowledge processes which Cope and Kalantzis (2016) identify as experiencing (the known and the new), conceptualizing (naming and use of theory), analyzing (functionally and critically), and applying (appropriately and creatively). Table 9.2 represents a sample of ways to map the multiliteracies framework which is Learning by Design into literacy objectives. It illustrates an example of how the Learning by Design framework can be mapped to curriculum objectives for all children. Thus, Table 9.2 is an example of how literacy curricula can be aligned with the Learning by Design framework so that the pedagogical elements can pair with the representational modes of meaning (oral, visual, tactile, etc.) upon which learners can draw and which educators can utilize.

Table 9.2.
Learning by Design Framework Mapped to Sample of Literacy Objectives in Ontario Grade 10 English Curriculum

Knowledge Process	Sample Literacy Objectives
Experiencing the known	Demonstrate understanding of content (Eng2D-1.3)
Experiencing the new	Extend understanding of texts (Eng2D-1.5)
Conceptualising by naming	Oral Communication (Eng2D-1.3)
Conceptualising by theory	Interpreting (Eng2D-1.5); Metacognition (Eng2D-3.1)
Analysing functionally	Reading for meaning (Making Inferences, meaning, organizing ideas) Eng2D-1.4); Extending understanding (Eng2D-1.5); Analysing texts (Eng2D-1.6)
Analysing critically	Demonstrate reflective practices (Evaluating texts - Eng2D-1.7, 1.8)
Applying appropriately	Reflect—Metacognition (Eng2D-4.1)
Applying creatively	Innovate—Interconnected skills (Eng2D-3.2, 4.2)

REFERENCES

Ajayi, L. (2010). Preservice teachers' knowledge, attitudes, and perception of their preparation to teach multiliteracies/multimodality. *The Teacher Educator*, *46*(1), 6–31. https://doi.org/10.1080/08878730.2010.488279

Ajayi, L. (2011). A multiliteracies pedagogy: Exploring semiotic possibilities of a Disney video in a third grade diverse classroom. *The Urban Review*, *43*(3), 396–413. https://doi.org/10.1007/s11256-010-0151-0

Alexander, J. (2009). Gaming, student literacies, and the composition classroom: Some possibilities for transformation. *College Composition and Communication*, *61*(1), 35–63.

Apperley, T., & Beavis, C. (2013). A model for critical games literacy. *E-learning and Digital Media*, *10*(1), 1–12. https://doi.org/10.2304/elea.2013.10.1.1

Apperley, T., & Beavis, C. (2011). Literacy into action: Digital games as action and text in the English and literacy classroom. *Pedagogies*, *6*(2), 130. https://doi.org/10.1080/1554480X.2011.554620

Beavis, C. (2012). Video games in the classroom: Developing digital literacies. *Practically Primary*, *17*(1), 17–20.

Brochu, P., Gluszynski, T., & Knighton, T. (2010). *Measuring up: Canadian results of the OECD PISA study: The performance of Canada's youth in reading, mathematics and science: 2009 first results for Canadians aged 15*. Ottawa, Canada: Human Resources and Social Development Canada & Statistics Canada. Retrieved from http://publications.gc.ca/collection_2010/statcan/81-590-X/81-590-x2010001-eng.pdf

Brochu, P., Deussing, M.-A., Houme, K., & Chuy, M. (2013). *Measuring up: Canadian results of the OECD PISA study. The performance of Canada's youth in mathematics, reading and science: 2012 first results for Canadians aged 15*. Toronto, Canada: Council of Ministers of Education, Canada. Retrieved from http://www.cmec.ca/252/Programs-and-Initiatives/Assessment/Programme-for-International-Student-Assessment-%28PISA%29/PISA-2012/index.html

Connell, R. (1996). Teaching the boys: New research on masculinity, and gender strategies for schools. *Teachers College Record*, *98*(2), 206–235.

Cope, B., & Kalantzis, M. (2009). "Multiliteracies": New literacies, new learning. *Pedagogies: An International Journal*, *4*(3), 164–195. https://doi.org/10.1080/15544800903076044

Cope, B., & Kalantzis, M. (2016). *A pedagogy of multiliteracies: Learning by design*. Basingstoke, England: Palgrave Macmillan.

Duncum, P. (2004). Visual culture isn't just visual: Multiliteracies, multimodality and meaning. *Studies in Art Education*, *45*(3), 252–264.

Fox, T. (2015). *Undertale* [Video game]. American Indie.

Frank, B., Kehler, M., Lovell, T., & Davison, K. (2003). A tangle of trouble: Boys, masculinity and schooling—future directions. *Educational Review*, *55*(2), 119–133. https://doi.org/10.1080/0013191032000072173

Freeman, A., Adams Becker, S., Cummins, M., Davis, A., & Hall Giesinger, C. (2017). *NMC/CoSN Horizon Report: 2017 K–12 Edition*. Austin, Texas: The New Media Consortium. Retrieved from https://cdn.nmc.org/media/2017-nmc-cosn-horizon-report-k12-EN.pdf

Ganapathy, M. (2014). Using multiliteracies to engage learners to produce learning. *International Journal of e-Education, e-Business, e-Management and e-Learning, 4*(6), 410–422. doi:10.17706/IJEEEE.2014.V4.355

Gee, J. P. (2003). *What video games have to teach us about learning and literacy* (1st ed.). New York, NY: Palgrave Macmillan.

Gee, J. P. (2007). *Good video games + good learning: Collected essays on video games, learning, and literacy.* New York, NY: Peter Lang.

Gee, J. P. (2008). *What video games have to teach us about learning and literacy.* New York, NY: Palgrave Macmillan.

Gee, J. P. (2014). *What video games have to teach us about learning and literacy.* New York, NY: Palgrave Macmillan.

Greig, C. (2003). Masculinities, reading and the "boy problem": A critique of Ontario policies. *EAF Journal, 17*(1), 33–56.

Gros, B. (2007). Digital games in education: The design of games-based learning environments. *Journal of Research on Technology in Education, 40*(1), 23–38.

HAL Laboratory. (2014). *Super Smash Bros.* Melee [Video game]. Kyoto, Japan: Nintendo.

Hommel, M. (2010). Video games and learning. *School Library Monthly, 26*(10), 37–40.

Jansz, J. (2005). The emotional appeal of violent video games for adolescent males. *Communication Theory, 15*(3), 219–241.

Jenkins, H. (2002). Game design as narrative architecture. In N. Wardrip-Fruin & P. Harrigan (Eds.), *First person: New media as story, performance, and game* (pp. 118–130). Cambridge, MA: MIT Press.

Jenkins, H. (2006). *Convergence culture: Where old and new media collide.* New York, NY: New York University Press.

Kafai, Y. B., & Burke, Q. (2016). *Connected gaming: What making video games can teach us about learning and literacy.* Cambridge MA: MIT Press.

Kalantzis, M., & Cope, B. (2012). *Literacies.* Cambridge, England: Cambridge University Press.

Kehler, M. D. (2007). Hallway fears and high school friendships: The complications of young men (re)negotiating heterosexualized identities. *Discourse: Studies in the Cultural Politics of Education, 28*(2), 259–277. https://doi.org/10.1080/01596300701289375

Kehler, M. D. (2010). Boys, books and homophobia: Exploring the practices and policies of masculinities in school. *McGill Journal of Education/Revue Des Sciences de L'éducation de McGill, 45*(3), 351–370.

Lingard, B., & Douglas, P. (1999). *Men engaging feminisms: Pro-feminism, backlashes and schooling.* Buckingham, England; Philadelphia, PA: Open University Press.

Lingard, B., Martino, W., Mills, M., & Bahr, M. (2002). *Addressing the educational needs of boys.* Research report submitted to the Department of Education, Science and Training, Australia. Retrieved from https://epublications.bond.edu.au/cgi/viewcontent.cgi?referer=https://www.google.com/&httpsredir=1&article=1001&context=mark_bahr

Lingard, B., Martino, W., & Mills, M. (2009). *Boys and schooling: Beyond structural reform.* Basingstoke, England: Palgrave Macmillan.

Mac an Ghaill, M. (1994). *The making of men: Masculinities, sexualities and schooling*. Buckingham, England: Open University Press.

Martino, W. J. (2013). On a commitment to gender and sexual minority justice: Personal and professional reflections on boys' education, masculinities and queer politics. In M. B. Weaver-Hightower & C. Skelton (Eds.), *Leaders in gender and education: Intellectual self-portraits* (pp. 163–177). Rotterdam, Netherlands: Sense.

Martino, W., & Rezai-Rashti, G. (2013). 'Gap talk' and the global rescaling of educational accountability in Canada. *Journal of Education Policy, 28*(5), 589–611. https://doi.org/10.1080/02680939.2013.767074

Mojang. (2011). Minecraft [Video game]. Stockholm, Sweden: Author.

National Center for Education Statistics. (2013). *The Nation's Report Card: Trends in academic progress 2012* (NCES 2013 456). Washington, DC: Institute of Education Sciences, Department of Education. Retrieved from https://nces.ed.gov/nationsreportcard/subject/publications/main2012/pdf/2013456.pdf

Naughty Dog. (2013). *The Last of Us* [Video game]. Santa Monica, CA: Sony Computer Entertainment.

Newkirk, T. (2002). *Misreading masculinity: Boys, literacy, and popular culture*. Portsmouth, NH: Heinemann.

New London Group. (1996). A pedagogy of multiliteracies: Designing social futures. *Harvard Educational Review, 66*(1), 60–93. https://doi.org/10.17763/haer.66.1.17370n67v22j160u

New London Group. (2000). A pedagogy of multiliteracies: Designing social futures. In B. Cope & M. Kalantzis (Eds.), *Multiliteracies: Literacy learning and the design of social futures* (pp. 9–37). London, UK: Routledge. Retrieved from: http://www.worldcat.org/title/multiliteracies-literacy-learning-and-the-design-of-social-futures/oclc/41400924/viewport

Organization for Economic Co-Operation and Development. (2013). *PISA 2012 results in focus: What 15-year-olds know and what they can do with what they know: Key results from PISA 2012*. Retrieved from http://www.oecd.org/pisa/keyfindings/pisa-2012-results.htm

Organization for Economic Co-Operation and Development. (2016). *PISA 2015 results: Excellence and equity in education* (Volume 1). Retrieved from https://www.cmec.ca/Publications/Lists/Publications/Attachments/365/PISA2015-CdnReport-EN.pdf

Ontario Ministry of Education. (2004). *Me read? No way! A practical guide to improving boys' literacy skills*. Toronto, Canada: Ministry of Education. Retrieved from http://www.edu.gov.on.ca/eng/teachers/publications.html#2004

Ontario Ministry of Education. (2009a). *Me read? And how! Ontario teachers report on how to improve boys' literacy skills*. Toronto, Canada: Ministry of Education. Retrieved from http://www.edu.gov.on.ca/eng/curriculum/boysliteracy.html

Ontario Ministry of Education. (2009b). *The road ahead—boys' literacy teacher inquiry project, 2005 to 2008, Final Report*. Toronto, Canada: Ministry of Education. Retrieved from http://www.edu.gov.on.ca/eng/curriculum/boysliteracy.html

Patton, M. Q. (2002). *Qualitative research and evaluation methods* (3rd ed.). Thousand Oaks, CA: SAGE.

Piaget, J. (1972/2008). Intellectual evolution from adolescence to adulthood. *Human Development, 51*(1), 40–47. doi:10.1159/000112531

Pillay, H. (2002). An investigation of cognitive processes engaged in by recreational computer game players: Implications for skills for the future. *Journal of Research on Technology in Education, 34*(3), 336–350. https://doi.org/10.108 0/15391523.2002.10782354

Rowsell, J., & Walsh, M. (2011). Rethinking literacy education in new times: Multimodality, multiliteracies, & new literacies. *Brock Education, 21*(1), 1. https:// doi.org/10.26522/brocked.v21i1.236

Sanford, K., & Madill, L. (2006). Resistance through video game play: It's a boy thing. *Canadian Journal of Education, 29*(1), 287–306, 344–345.

Sanford, K., & Madill, L. (2007). Understanding the power of new literacies through video game play and design. *Canadian Journal of Education, 30*(2), 432–455.

Skerrett, A. (2011). "Wide open to rap, tagging, and real life": Preparing teachers for multiliteracies pedagogy. *Pedagogies, 6*(3), 185. https://doi.org/10.1080/1 554480X.2011.579048

Squire, K. D. (2013). Video game-based learning: An emerging paradigm for instruction. *Performance Improvement Quarterly, 26*(1), 101–130.

Steinkuehler, C. (2007). Massively multiplayer online gaming as a constellation of literacy practices. *E-Learning and Digital Media, 4*(3), 297–318. http://dx.doi. org/10.2304/elea.2007.4.3.297

Steinkuehler, C. (2011). *The mismeasure of boys: Reading and online videogames.* Madison, WI: Wisconsin Center for Education Research, University of Wisconsin.

Steinkuehler, C., Squire, K., & Barab, S. A. (2012). *Games, learning, and society: Learning and meaning in the digital age.* Cambridge, England: Cambridge University Press.

Ubisoft Montpellier. (2014). *Valiant hearts: The great war* [Video game]. Carentoir, France: Ubisoft.

Upper One Games. (2014). *Never alone* (Kisima Innitchuna) [Video game]. Anchorage, AK: E-Line Media.

Valve Corporation. (2011). *Portal 2* [Video game]. Bellevue, WA: Author.

VanSledright, B. A. (2002). Fifth graders investigating history in the classroom: Results from a researcher-practitioner design experiment. *The Elementary School Journal, 103*(2), 131–160. https://doi.org/10.1086/499720

Vygotsky, L. S. (1978). Interaction between learning and development. In M. Cole, V. John-Steiner, S. Scribner, & E. Souberman (Eds.), *Mind and society: The development of higher psychological processes* (pp. 79–91). Cambridge, MA: Harvard University Press. (Original work published 1935)

Wenger, E. (1998). *Communities of practice: Learning, meaning and identity.* Cambridge, England: Cambridge University Press.

Wolcott, H. F. (1987). On ethnographic intent. In G. Spindler & L. Spindler (Eds.), *Interpretive ethnography on education: At home and abroad* (pp. 37–51). Hillsdale, NJ: Erlbaum.

PART IV

DESIGN AND BENEFITS OF
SCHOOL-UNIVERSITY-COLLABORATIONS

CHAPTER 10

DIGITAL SCHOOL NETWORKS

Technology Integration as a Joint Research and Development Effort

Michael Kerres and Bettina Waffner
University of Duisburg-Essen, Germany

Schools are increasingly challenged to develop a digital strategy. For some time, digital technology in classrooms has been a topic only for few teachers but, increasingly, it is perceived as a strategic issue for schools that impacts the organization more deeply. The topic not only relates to the deployment of technology and infrastructure and the training of teachers; schools must also address the consequences and potential of technology for curriculum reform and new instructional methods, which hints at a larger project of school development.

In order to understand the transformative power of digital technology and the developmental options for schools, a research and development program brings together K–12 schools within a geographical region to jointly discuss and develop their digital strategies under the support of a university's learning lab. This chapter describes the rationale of the school-

Integrating Digital Technology in Education:
School–University–Community Collaboration, pp. 227–241
Copyright © 2019 by Information Age Publishing
All rights of reproduction in any form reserved.

university-community collaborative educational research effort and reflects on the role of the university's learning lab in this development process.

REVISITING THE DIFFERENCE BETWEEN EDUCATION AND RESEARCH ON EDUCATION

Scholarly studies in the field of education are expected to solve current challenges in society. These works in education relate to various other domains of research such as psychology, sociology or philosophy; in fact, they rely on results and methods defined by other disciplines. Nevertheless, research in the field of education must be different than other strands of scientific effort if it wants to meet society's expectations. Ever since education was emancipated from philosophy and established as a discipline of its own, the question has been raised of how to conceptualize research in education next to the closely related scholarly disciplines. The inaugural issue of *Educational Review* was published in 1891 with a paper titled "Is there a science of education?" by Harvard philosopher Josiah Royce (Lagemann, 2002). For a long time, educational research confined itself to the practice of applying knowledge from other fields to the field of education. Gradually, some researchers followed the logic and methods established in other disciplines, such as educational psychology, which was a successful strategy in the academic world. However, it became quite obvious that this approach—despite being successful for the individual researcher—would not be able to address the challenges that society was facing in the field of education. For example, even today large-scale assessments in education, such as the Program for International Student Assessment (PISA), have garnered much public attention and have offered important snapshots into the state of education in different countries around the world but they typically fail to provide insights into what changes are needed and how these changes in education might be achieved (Meyer & Benavot, 2013).

In his systems theory of society, Luhmann (1995) described how modern societies have developed functionally separate, loosely coupled (Weick, 1976) subsystems in law, education, health, politics, science, and so on. The agenda of each subsystem follows a different rationale and it is not obvious how the value created in one subsystem can be transferred beyond its boundaries. Therefore, scientific research essentially has to follow its own agenda if it does not want to be subsumed as a part of the educational subsystem. According to this view of systems theory, there is an essential difference in the aims and rationale of the field of education on the one hand and research on education on the other. The implications of Luhmann's view for research on education have been widely discussed and give reason for a somewhat pessimistic view on the potential of educational research for improving educational practice and contributing to educational reform. Markauskaite, Freebody, and Irwin (2011) described various approaches that have been developed to overcome the outlined barrier between research and practice in the social science, such as *action research* (Groundwater-Smith & Irwin, 2011) or *design-based research* (Brown, 1992; Sandoval & Bell, 2004). Other recent approaches lean more towards practice and the analysis of professional expertise, such as the *scholarship of teaching* (Hutchings & Shulman, 1999) and the *teacher as researcher* movement that emphasizes practitioner reflection on personal routines and experiences (Cochran-Smith & Lytle, 1999).

THE LEARNING LAB'S APPROACH: JOINT RESEARCH AND DEVELOPMENT IN DIGITAL SCHOOL NETWORKS

At the University of Duisburg-Essen (Germany), the learning lab is a unit for innovation and design-based research in the different sectors of education. With its mission "exploring the future of learning," the lab cooperates with various schools in the region of North-Rhine Westphalia to develop new perspectives for learning and teaching with digital media. These projects vividly demonstrate the discrepancy of the aims of the educational practice on the one hand and the aims of educational research on the other hand. Schools, for their part, need to develop a digital strategy and want to gain knowledge about the implications of digitization for teaching and learning. They want to draw from the experiences and results of research, which the university can provide. Also, schools are interested in the learning lab's expertise in developing and managing change in the field of educational innovations. In sum, they perceive the university as a consultant in organizing and supporting the learning processes of educational organizations. At the same time, the university wants to gain general knowledge about how schools develop digital strategies, how they organize the digital transformation of education, and what key factors contribute to success or failure. Various models and concepts describe the process of digital technology integration in schools (Davies & West, 2014; Hall & Hord, 2001; Owston, 2007; Petko, Egger, Cantieni, & Wespi, 2015); these models guide the theorizing and the development of hypotheses for nascent innovation. One basic research question considers a hypothetical model of stepwise progression for educational institutions in the adoption of technology from attention to sustainable anchoring.

The "NMC Horizon Report" by Freeman, Adams Becker, Cummins, Davis, and Hall Giesinger (2017) described the 5-year impact of innovative practices and technologies for primary and secondary education. The report highlighted several key trends, significant challenges, and developments in educational technology that have influenced teaching, learning, and creative inquiry in K–12 education. Some key trends reflect the growth in technology and its impact on education while others relate to teaching practices, the role of teachers, and evolving structures and processes in schools. It is interesting to note that these more global topics have gained recognition in the "NMC Horizon Reports" in recent years. Digital technology is no longer a topic that is confined to the discussion of a few experts and some pioneering teachers interested in the use of technology for instructional reform. Digital technology has reached mainstream discourse; educators acknowledge that schools are affected at an organizational level and it is no longer the individual teacher alone who will reap the potential benefits of digital technology for innovations in teaching and learning. Learning lab's approach of *digital school development* illustrates the new attention to trends and challenges on the organizational level and how schools can adopt digitization as an organization. This directly links to "Rethinking How Schools Work" and "Rethinking the Roles of Educators," which were identified as key trends by Freeman et al. (2017).

Currently, schools are in the midst of the digital transformation of education. What this will be like is not determined solely by digital technology, but rather is a question of design in which practices and standards are negotiated locally in organizations as well as on a broader social level. These processes of transition offer many opportunities for innovation but present great challenges and uncertainties at the same time. Schools are confronted with the task of designing teaching and learning environments for a digitized world while considering the instructional potential of digital media, thereby addressing their educational mission to enable students to participate in society and culture. In the past, schools made early attempts to acquire digital technology and to train teachers accordingly. However, these efforts proved unsatisfactory because established structures and processes as well as traditional teaching and learning practices were not successively adapted to the changed conditions. A crucial element of German schools as a system is their stability, which is characterized by the Prussian military tradition. Characteristics of this tradition include a strict hierarchical structure, discipline, order, and traditions. Thus, a culture of openness is uncommon in the German educational sector; similarly, a trial-and-error practice, which views mistakes as opportunities instead of failures, is rare. In order to understand the transformative power of digital technology and the developmental options for schools, the German regional school boards initiated a research and development program in 2016 where university researchers cooperate with local schools. The basic idea was to bring together schools within a geographical region to jointly discuss and develop their digital strategies under the support of the university's learning lab.

The research literature offers various models to describe the digital transformation of schools. Some models focus on the fields of actions necessary to address the various challenges (Cabrol & Severin, 2009). Other models describe the process of school development as a sequence of steps, such substitution, augmentation, modification, and redefinition (Puentedura, 2012). The TPCK-model (Angeli & Valanides, 2009) describes teacher competencies necessary for integrating technology in classrooms: In addition to content knowledge (CK) on a given subject and pedagogical knowledge (P), a competent teacher will need technological knowledge (T) as well. This model describes the development of teaching competences at the intersection of each element (e.g., how technology knowledge is necessary or contributing to content knowledge).

These theoretical models have been consulted as a backdrop for the *digital school networks* where schools collaboratively develop strategies and measures. The primary research question is how these models can be used as an input for school consultation; second, we want to know how accurately these models describe the challenges and processes schools face in their endeavors to implement digital strategies. After all, decision-making in education policy is complex and the process of organizational change as well as the relationship between science and practice are even more complex. Specifically, this chapter will examine this last argument in more detail from a system-theoretical perspective.

HOW TO...? DIGITAL SCHOOL NETWORKS IN PRACTICE

As described above, educational practice and educational research are two communities that generate and use knowledge in different ways. They each follow a different logic in goals, interests, working methods, and basic assumptions. This two-community-metaphor (Farley-Ripple, May, Karpyn, Tilley, & McDonough, 2018) highlights the challenge of overcoming research/practice incongruence in order to track evidence-based change processes in school practices. This is the task of the learning lab as a university partner in digital school networks. The practice community consists of committed teachers, school principals, stakeholders in state teacher training, and school authorities at the local political level. As a scientific community, learning lab establishes, accompanies, and organizes the process of digitizing teaching and learning formats in schools with the aim of achieving a fruitful cooperation between the research and practice communities. Learning lab's thesis is that these processes (establishing, accompanying, and organizing) will be successful by bridging the gap between the two different communities through points of contact and overlaps, which can lead to an improvement in practice based on empirically verified current research results.

The German educational landscape has a specific federal structure. The majority of schools is public, where 16 different federal state ministries determine the curriculum and structure. At the municipal level, the school authorities are responsible for the technical equipment. At all political levels, science and research are expected to provide evidence-based decision-making assistance on educational policies. Politicians and school practitioners ask researchers to identify the measures that best fulfil the educational mandate in the digital world; the public then expects teachers to implement measures in order to have the desired effect (Farley-Ripple et al., 2018). However, in times of change and upheaval, this arrangement is understandable but is too one-dimensional. The implicit political assumptions behind policy-making decisions reveal why the process cannot have a successful outcome in this way. Decisions on educational policy are not solely determined by scientific expertise but are a complex phenomenon as political science has demonstrated since the 1960s (Easton, 1965). Political institutions can be understood as systems in exchange with their environment. Demands by singular groups of citizens, associations, or political advisories constitute the input for the policy decisions in a complicated process. The overarching goal is to provide either conservation or change in societal conditions (Nullmeier & Wiesner, 2003). This systems theory approach led to the development of the policy cycle model with six phases within the complex political decision-making process (Windhoff-Heritier, 1987). Furthermore, the policy process does not end with the implementation of policy outputs. Thus, there is no direct connection between the scientific and the practical communities to implement current research results.

Regional or community boards, which are responsible for financing all school infrastructure (buildings, furniture, electricity, computer equipment, networks, etc.), are the sponsors for the digital school networks; meanwhile, state authorities pay the teachers. With learning lab's support, 70 schools in six networks have been working together for 1 to 2 years. Each network meets in different settings with committees gathering in meetings, workshops, or barcamps. Stakeholder attend as they are affected by the outcomes. Thus, principals might discuss topics around the level of organizational development, while teachers might meet to discuss topics concerning pedagogical and didactical questions and the political leadership group might meet to address process management. The learning lab coordinates and supports all events. This is how coherence and continuity are addressed and supported within the school network. These face-to-face meetings are crucially important for the collaboration of the stakeholders.

This is the reason why school networks are geographically close with short distances to each meeting.

The participating schools are public and are not competitors in their region. Therefore, collaboration within a network is unaffected by inter-regional rivalries. At the same time, the learning lab tries to condense generalizable knowledge on how schools actively cope with the process of digital transformation. In this way, the learning lab founds, accompanies, and supports the digitization of teaching and learning in all schools with the aim of achieving fruitful collaboration and bridging the gap between research and practice. The schools and university maximize mutual points of contact and overlaps, which leads to an improvement in practice based on empirically verified current research results.

From a system-theoretical perspective, theory and practice usually refer to interdependent, complementary elements of a change process in which the practice community can best identify the problems to be solved (Cohen & Manion, 2018). Therefore, the learning lab designed the initial entry into cooperative networking as a workshop, divided into three phases, which aimed to achieve a common vision for future schools. Led by the learning lab, the practice community identifies problems and challenges of working with digital media both currently and in an increasingly digitally-shaped world. This phase contrasts with a second phase where the team takes up the negative challenges identified in phase one and turns them into positive benefits in an imaginary utopian world. This can be achieved by the following steps: Participants imagine a world where there are no administrative, technical, or organizational restrictions. What would such a world look like when integrating digital media? If the restrictions were not present anymore, where would there be opportunities in their work? Within this phase, participants are often able to let go of familiar but hindering doubts.

Participants are empowered to develop a set of goals independent of the obstacles that inherently adhere to such considerations. Subsequently, the team creates an overarching vision of learning, teaching, and working with digital media at the school. This process allows for the identification of problematic areas and the development of initial explanations and hypotheses, which form the basis for further joint work.

In this context, it is important to point out that the learning lab has a specific understanding of the term *problem*. Problems (and conflicts) are not destructive per se; they contain a high potential energy, which can be the starting point for initiating change. This is a highly sophisticated procedure, which requires professional organization of the steps. When discussing an identified problem, a dynamic process develops between stakeholders, which they can use fir a systematic, empathic, and above all analytical support of constructive dialogue. By risking disturbances and irritations, participants can call into question supposed certainties, which allows for reflection. The process is worthwhile for developing innovative solution strategies. For example, a common argument among many teachers and principals is that technical equipment must first be available before approaching the topic of learning within a digitized world. When learning lab provocatively states that this is a welcome argument for those who do not want to face a necessary change process, stakeholders are often irritated or defensive. After this direct exposure, it is possible to discuss the apprehensions in the context of change. Examples of apprehensions include fear of excessive demands, anxiety that one's lack of knowledge might be exposed, or simply the lack of time. If the learning lab takes these worries seriously, then the topic of digital education can be addressed on a new level. The learning lab carries out this work as a research community. The scientific analysis of the dynamics and practices against the backdrop of theory can provide deeper understanding of the change processes (McNiff, 2010). In analyzing the problems, the learning lab considers state-of-the-art research, which allows a view of the problem areas through diverse perspectives.

Problems can be addressed and solved using this reflexive process. Collaborative thinking and working are central building blocks of working in the learning lab's school networks. However, this does not mean that all actors work equally intensively on all aspects of problem solving; rather they develop measures and strategies in their respective, specific, thematic areas. The learning lab understands this as an organizational development process and identifies significant fields of action with the respective key stakeholders in order to structure the work in the network. In addition to technology and personnel development, which have already been described above as fields of action, the network must also pay attention to curriculum development and structures of organization in the change process in order for learning and teaching in the digital era to be effective and professional.

Joint work with the practice community follows this analytical work. Before taking any steps to overcome challenges, the learning lab helps identify the implicit assumptions or hidden agendas (such as attitudes, values, goals, etc.), which accompanied the identification of challenges. This is possible with the help of sound analysis based on research findings that can be transferred into the practice community. This work offers the opportunity to modify the former critical standpoints. In steering groups, school authorities and, if applicable, other stakeholders agree on schedule of topics and dates for network meetings with the learning lab. The emphasis is on persistent, joint, network meetings conceived methodically as a *barcamp*. The intensive dialogue that takes place across school borders is an opportunity to develop problem-solving strategies. The organizers often form specific, topic-related working groups within network meetings in order to work on individual ideas in greater depth.

After schools and school authorities of a given region have made the decision to understand media integration as a lengthy and complex organizational development process in which joint work in a school network under the leadership of the scientific expertise of the university is deemed helpful, a so-called duty-stapler serves to cement the joint agreement. The work in the network is long-term and requires mutual, reliable support. In this respect, it is important to engage in joint work reliably and permanently.

Networking begins with the development of a common vision that focuses on the following questions: What can school life and learning in a digitized world look like? Which rules for handling of digital media should we follow? What should learning and teaching look like? A media development process can trigger these questions. The gradual development of new and innovative ideas (that consider the advantages of using digital media such as more flexible learning arrangements or spatial independence) is also effective in pulling extant thought patterns and school practices into the light. In addition, digital media potential develops only if technology is in the hands of learners, which allows for searches of educational resources in mediums such as text, film, sound, or picture, which can be compiled collaboratively (i.e., working on a subject in ePort). Analysis of the common vision of the participants in a given network makes it possible to identify fields of action, which form the basis for further joint work.

A steering committee is responsible for planning implementation of the goals (decided upon during the kick-off event in the different fields of action). The steering committee includes members of the school board, actors of the federal state ministry, municipal politicians, school principals, and the learning lab. If necessary, other supporting stakeholders such as municipal media advisors, school authorities, or teacher training experts can also join the committee, which meets at least quarterly at the invitation of the school authority.

The school management group coordinates the timing of the joint work (i.e., planned training measures). This committee, which in addition to the school authorities of the participating schools and the learning lab also includes the local school authorities and any other actors such as school media representatives, discusses and coordinates the administrative aspects of digital media integration. The network meetings are an important element and open to teachers and headmasters equally. In the form of a barcamp, an intensive collegial exchange takes place on topics that the teachers themselves bring in and share. In this way, the group is able to use existing practical knowledge and discuss topics that are currently relevant for schools.

Training courses on specific topics, in depth, are another aspect of the network meetings. The learning lab is responsible for the concept and organization; one of the central principles is to generate experts from the network itself. Only in exceptional cases do external partners support the exchange of best practices. In early network meetings, competition between school administrators from participating schools is still noticeable. Consequently, the learning lab must build trust among schools, which is a central requirement for working together. This illustrates the particular importance of a long-term process that requires special support and moderation. The establishment of cooperative and collaborative forms of further training and task work means nothing less than a paradigm shift.

In the beginning, the capacities and goals of the municipalities currently supported by learning lab are very different. The various networks differ according to their location (situated in an urban or rural context), representativeness (the proportion of schools involved compared to the total number of schools in a municipality), and size (how large each network is overall). Learning lab currently supports school networks of five to 16 participating schools. Paying attention to this heterogeneity is central for success; the learning lab shapes the organization of each network differently with regard to locations and access routes as well as the coordination processes. The respective commitment of the network partners and the possibility of involving supporting stakeholders is also different. Each network has individual structures and work plans with the processes adapted to the needs of the project partners in regular meetings.

The experiences of the learning lab in one setting can also be of advantage in other networks. Thus, different methods such as survey settings, concepts, and practical knowledge for teaching and learning with digital media can be transferred. In this way, the networks benefit from each other as the learning lab brings together different experiences. However, it is not only the structures and organizations of the respective networks that are different. The content objectives may also differ, as they are determined, for example, by the existing technical equipment or the commitment and interest of teachers. Some networks work on basic technical equipment, while others tend to focus on the implementation of collaborative and cooperating forms of work. In this way, the university partner has a decisive structuring role in the network, which distinguishes the form of committee work from the everyday actions of school stakeholders. Operational thinking often prevails in practical work but in this process it is also worthwhile to think strategically with long-range targets in mind. The school network offers the possibility to depart from familiar modes of operation and to address problems on a different level. Promoting and fostering this form of thinking and working is an essential task of the learning lab. Practices, the understanding of processes, and conditions in which problems are situated are discussed and reflected on the basis of content-related impulses from the research community.

BRIDGING THE GAP TOWARDS A COMMON GOAL: LEARNING IN AND FOR A DIGITIZED WORLD

The learning lab grounds our work in school networks on an approach that is often called action research in educational science. Action research aims at a deeper understanding of the complexity of teaching and learning against the backdrop of social change processes. The method is characterized by a democratic, collaborative approach, in which a free flow of information exists, overcoming formal boundaries, and often takes place in informal settings. What is important here is the ideal communication identified by Habermas (2011) in which the power of the better argument works. The structure of school organizations is hierarchical, formal, and bureaucratic, while the working principles of school networks are contrary and characterized by collegiality, collaboration, and openness. The contiguity can create tension. The cooperation between schools and universities offers potential for interplay. If this collaborative way of working and thinking is reflected upon by an individual and new ground is broken by following new ways of working, then practice becomes more effective and professional. Kemmis (2009) describes this process as "practice-changing practice" (p. 464). Action research encourages a variety of interpretations and approaches, which should not lead to a uniform solution but rather to quite different, individual results for individual schools. In this way, a creative process of plurality can emerge.

In previous years, the relation between science and practice has often been the subject of scientific research in order to identify key success factors for their cooperation and to clarify how current scientific research results can be given practical relevance (for a deeper understanding, Coburn & Penuel, 2016). As a scientific partner, the learning lab has been gathering practical experience in cooperation with school partners for several years. However, so far, researchers have paid little attention to a crucial issue: How are practice, science, and evidence-based research results related to each other? It is particularly interesting to see how the research community generates knowledge and how it is used by the practicing community. How do the two communities influence each other? And how do they influence educational policy decision-making?

The scientific approach of the learning lab is founded on design-oriented educational research. From this perspective, we see no value per se in digitization and mediatization of society but there is some potential to solve today's educational problems and concerns. The quality of innovative learning arrangements can only be measured through the concrete application of them. In this respect, practical application is of immanent importance for the learning lab's scientific work, which does not take place through external observation but is based on the equal work of two different, complementary partners towards a common goal: developing innovative teaching and learning in and for a digitized world. This project will only succeed if the gap between the practice and research communities can be bridged and they become a school-university-community in which various rationales of the individual partners have their justified place. The work of the learning lab in the digital school networks offers a suitable field of research to investigate the role of Luhmann's (1995) subsystems scientifically, due to the already existing relationship of trust between these actors.

The analytical framework offered by Farley-Ripple et al. (2018) is a suitable tool for an initial approach to the subject area. Along two dimensions, the framework helps identify structures, basic assumptions, perspectives, and processes that operate below the surface within the research and practice communities. One dimension identifies the depth with which processes are changed and traditional basic assumptions are recognized and reflected upon (Coburn, 2003). The second dimension identifies the gap between the two communities in regards to their outlook and attitudes towards research. Farley-Ripple et al. assumed that size of the research/practice gap would be inversely proportional to the depth of process change. In other words, bridging the gap could mean intense depth for transforming learning and teaching.

FINAL REMARKS

In our work on digital school development, we are directly confronted with the problem of how to combine the demands of the practical field with the agenda of the scientific world. Both fields do have research questions and, at first glance, these questions seem to be the same: How can we transform school for a digital world? How can we prepare teachers and students for education in a digital world? How can we use digital technology to enhance teaching and learning? What are the best digital tools and platforms for teaching and learning and how can these tools be introduced?

Teachers seek solutions to their developmental challenges that fit the scenario in which they work. They need solutions that fit their situation and yet every school is different. Therefore, answers from one school cannot easily be transferred to other schools. In a traditional approach, scholarly researchers would systematically maintain some distance as they observe, document, categorize, systematize, and analyze the answers schools have found. In contrast, using a dialogical understanding of design-based research, researchers would actively participate in developing solutions with schools, letting teachers gain from the researchers' prior experiences with other schools and the expertise of documented experiences in the research literature. In the latter case, researchers must reflect on the dilemma that they are coproducer of solutions that have emerged in this process. Therefore, the achievement of digital school development could be due to the mere fact that the external experts have proposed (or inflicted) a certain model that structures the process. Thus, the researcher is confronted with an empirical reality of one's own making. However, there are several tools that can help to reduce this problem, which is common to qualitative research. In our case, documentation and reflection seem to be the most important and straight-forward solutions. Possible findings should be documented early and presented as preliminary findings in the ongoing dialogue. A process of validating findings from one setting with other schools is also another step in the continual improvement of theories and model-based research which can yield information for other schools. From our experience, this process of dialogical research in design projects needs further refinement. The aim should be to refine the methodology of such approaches where researchers and practitioners in education actively construct designs and environments for learning.

REFERENCES

Angeli, C., & Valanides, N. (2009). Epistemological and methodological issues for the conceptualization, development, and assessment of ICT–TPCK: Advances in technological pedagogical content knowledge (TPCK). *Computers & education, 52*(1), 154–168.

Brown, A. (1992). Design experiments: Theoretical and methodological challenges in creating complex interventions in classroom settings. *Journal of the Learning Sciences, 2*(2), 141–178. doi.org/10.1207/s15327809jls0202_2

Cabrol, M., & Severin, E. (2009). ICT to improve quality in education–A conceptual framework and indicators in the use of information communication technology for education (ICT4E). In F. Scheuermann & F. Pedró (Eds.), *Assessing the effects of ICT in education* (pp. 83–106). Luxembourg: European Union.

Cochran-Smith, M., & Lytle, S. L. (1999). The teacher research movement: A decade later. *Educational Researcher, 28*(7), 15–25. doi.org/10.3102/0013189X028007015

Coburn, C. E. (2003). Rethinking scale: Moving beyond numbers to deep and lasting change. *Educational Researcher, 32*(6), 3–12. doi:org/10.3102/0013189X032006003

Coburn, C. E., & Penuel, W. R. (2016). Research-practice partnerships in Education: Outcomes, dynamics, and open questions. *Educational Researcher, 45*(1), 48–54. doi.org/10.3102/0013189X16631750

Cohen, L., & Manion, L. (2018). *Research methods in education* (8th ed.). New York, NY: Routledge.

Davies, R. S., & West, R. E. (2014). Technology integration in schools. In J. M. Spector, M. D. Merrill, J. Elen, & M. J. Bishop (Eds.), *Handbook of research on educational communications and technology* (pp. 841–853). New York, NY: Springer.

Easton, D. (1965). *A system analysis of political life*. New York, NY: John Wiley & Sons.

Farley-Ripple, E., May, H., Karpyn, A., Tilley, K., & McDonough, K. (2018). Rethinking connections between research and practice in education: A conceptual framework. *Educational Researcher, 47*(4), 235–245. doi.org/10.3102/0013189X18761042

Freeman, A., Adams Becker, S., Cummins, M., Davis, A., & Hall Giesinger, C. (2017). *NMC/CoSN horizon report: 2017 K–12 edition*. Austin, TX: The New Media Consortium.

Groundwater-Smith S., & Irwin J. (2011). Action research in education and social work. In L. Markauskaite et al. (Eds.), *Methodological choice and design. Scholarship, policy and practice in social and educational research* (pp. 57–69). Dordrecht, the Netherlands: Springer.

Habermas, J. (2011). *Theorie des kommunikativen Handelns* (8th edn). Berlin: Suhrkamp.

Hall, G. E., & Hord, S. M. (2001). *Implementing change: Principles, patterns, and potholes*. Boston, MA: Allyn and Bacon.

Hutchings, P., & Shulman, L. S. (1999). The scholarship of teaching: New elaborations, new developments. *Change, 31*(5), 10–15. Retrieved from https://www.itl.usyd.edu.au/cms/files/Hutchings%20and%20Shulman.pdf.

Kemmis, S. (2009). Action research as a practice-based practice. *Educational Action Research, 17*(3), 463–474. doi.org/10.1080/09650790903093284

Lagemann, E. C. (2002). *An elusive science: The troubling history of education research*. Chicago, IL: University of Chicago Press.

Luhmann, N. (1995). *Social systems*. Stanford, CA: Stanford University Press.

Markauskaite, L., Freebody, P., & Irwin, J. (2011). *Methodological choice and design. Scholarship, policy and practice in social and educational research*. Dordrecht, the Netherlands: Springer.

McNiff, J. (2010). *Action research for professional development: Concise for new and experienced action researchers*. Dorset, England: September Books.

Meyer, H. D., & Benavot, A. (2013). *PISA, power, and policy: The emergence of global educational governance*. Oxford, England: Symposium Books.

Nullmeier, F., & Wiesner, A. (2003). Policy-forschung und verwaltungswissenschaft. In H. Münkler (Eds.), *Politikwissenschaft. Ein Grundkurs* (pp. 285–323). Hamburg, Germany: Rowohlt.

Owston, R. (2007). Contextual factors that sustain innovative pedagogical practice using technology: An international study. *Journal of Educational Change, 8*(1), 61–77. doi:10.1007/s10833-006-9006-6.

Petko, D., Egger, N., Cantieni, A., & Wespi, B. (2015). Digital media adoption in schools: Bottom-up, top-down, complementary or optional? *Computers & Education, 84*, 49–61. doi.org/10.1016/j.compedu.2014.12.019

Puentedura, R. R. (2012). *The SAMR model: Background and exemplars*. Retrieved from http://hippasus.com/resources/tte/

Sandoval, W. A., & Bell, P. (2004). Design-based research methods for studying learning in context: Introduction. *Educational Psychologist, 39*(4), 199–201. doi.10.1207/s15326985ep3904_1

Weick, K. E. (1976). Educational organizations as loosely coupled systems. *Administrative Science Quarterly, 21*(1), 1–19. doi:10.2307/2391875.

Windhoff-Heritier, A. (1987). *Policy-analyse. Eine Einführung*. Frankfurt, Germany: Campus Verlag.

CHAPTER 11

MUTUAL BENEFITS OF PARTNERSHIPS AMONG K–12 SCHOOLS, UNIVERSITIES, AND COMMUNITIES TO INCORPORATE A COMPUTATIONAL THINKING PEDAGOGY IN K–12 EDUCATION

Ahlam Lee
Xavier University

Computational thinking ability is necessary for students preparing to enter the 21st century workforce (Wing, 2006). However, general public stakeholders such as parents, K–12 community members, and policy makers do not understand what computational thinking constitutes, nor what its influence is on students' career readiness, which is detrimental to establishing a general consensus on the action necessary to incorporate a K–12 computational

Integrating Digital Technology in Education:
School–University–Community Collaboration, pp. 243–269

thinking pedagogy. Thus, such key stakeholders as university faculty or researchers in STEM, who might otherwise contribute to the development of a computational thinking pedagogy, are not aware fully of how such pedagogy can be beneficial for their own work. Accordingly, this chapter addressed the question, "What are the mutual benefits of establishing research partnerships among K–12 schools, universities, and communities to introduce a computational thinking pedagogy in K–12 education?" This question was guided by an Interest and Action framework (Thompson, Martinez, Clinton, & Díaz, 2017), which posits that research partnerships can be developed based on each stakeholder's *interest,* serve as a tool to take *action* to apply research to practice in education, and each stakeholder's interest and action arise from the potential benefits through building partnerships. A snapshot of each stakeholder's potential benefits is: (1) K–12 schools can cultivate the technology-rich learning environment necessary for all 21st learners; (2) universities can advance knowledge through research activities and have an opportunity to recruit prospective college students, and (3) such community members as industry leaders can develop new services or products related to a technology-embedded curriculum.

INTRODUCTION

In a rapidly growing digital world, a basic level of computational thinking (CT) is a core qualification for most professional occupations (Hoffmann, 2012), suggesting that all students should develop at least basic CT to be competitive in the 21st century workforce. CT constitutes the knowledge and skills of building computing systems and algorithmic approaches that are used to solve real-world problems (National Education Association, n.d.). Further, the U.S. Bureau of Labor Statistics (2017) projected that occupations requiring CT will grow by 19% from 2016 to 2026, the fastest growth rate among all occupations. Yet, the current K–12 curriculum does not fully support the development of CT (Grover & Pea, 2013). Many stakeholders are not fully aware of the significant role of CT in preparing a competitive workforce, which is detrimental to establishing consensus and action necessary to incorporate computational thinking pedagogy in K–12 education (Lee, 2015a; Wang, Hong, Ravitz, & Moghadam, 2016). As such, the first section of this chapter discusses the concept of CT and the current status on the adoption of CT pedagogy in classrooms at the K–12 level. The second section articulates how diverse stakeholders could benefit through building research partnerships to incorporate CT pedagogy in K–12 education. A core condition for building a research partnership with diverse stakeholders is to meet each stakeholder's self-interest in the partnership (Thompson et al., 2017), so it is important to discuss the mutual benefits for all.

CT PEDAGOGY IN CLASSROOMS AT THE K–12 LEVEL

The idea of computing education at the K–12 level has been around since the 1980s, when Seymour Papert coined the term *computational thinking* and developed LOGO programming for advancing children's procedural thinking (Grover & Pea, 2013). The growing interest in computing education was sparked when Jeannette Wing (2006) argued that CT is a core skill for students to enter the 21st century workforce. Now, the Common Core State Standards (CCSS) and Next Generation Science Standards (NGSS) place great emphasis on a more rigorous curriculum pertaining to critical thinking, conceptual understanding in math, and computer science (CS) all of which calls for the integration of CT pedagogy into K–12 classrooms, specifically in STEM classrooms (Tran, 2018). Nevertheless, computing education has been marginalized at the K–12 level (Grover & Pea, 2013; Tran, 2018). The underlying reasons for this marginalization includes (1) widespread misconceptions of CT detract from a compelling rationale for establishing facilities and training qualified teachers to implement CT pedagogy at the K–12 level (Topper & Lancaster, 2013); and (2) students are unmotivated to learn about CT-related concepts that are not directly tied to Advanced Placement (AP) or International Baccalaureate (IB) tests in such mainstream STEM subjects as sciences and math (Gallagher, Coon, Donley, Scott, & Goldberg, 2011). The next section provides a synthesis of the documented literature regarding what CT pedagogy teaches school-age students, what the research tells us about the learning process or outcomes of such pedagogy, and the practical challenges to implementing this pedagogy in K–12 classrooms.

Concept of CT Pedagogy

Policy makers, school administrators, and parents have a lack of understanding of how students learn CT. Indeed, in a nationally representative survey of multiple stakeholders in the K–12 community, the majority perceived that CT refers to simple point-and-click computer literacy skills such as creating Word documents or PowerPoint slides and using the Internet (Wang et al., 2016). Contrary to this prevalent misconception, CT pedagogy aims to develop students' knowledge and skills in building computing systems and algorithmic approaches rather than simply using built-in software like Excel or Word (National Education Association, n.d.). As such, CT could play a substantial role in solving real-world problems, including the prevention or cure of diseases and reduction of world hunger, through discovering a computing-based mechanism (Barr & Stephenson, 2011).

In this regard, CT pedagogy in K–12 classrooms is designed to provide students with a diverse set of such logical thinking and problem solving skills as abstraction, automation, and the development of algorithms for the construction of computing models or applications (Barr & Stephenson, 2011). As addressed earlier, however, CT pedagogy is a new concept and not commonly implemented in K–12 classrooms at the present time. Until recently, there was a substantial decline in schools offering CS courses that provide students with the fundamental knowledge and skills related to CT (Lee, 2015a; Wilson, Sudol, Stephenson, & Stehlik, 2010); however, since January 2018, 20 states have passed new laws or initiatives supporting CS at the high school level according to the Code.org Advocacy Coalition (Tech & Learning, 2018). Still, only a small number of high school students across the nation have the opportunity to take a CS course. Further, curriculum for computing education in K–12 classrooms is still underdeveloped, and there is a lack of qualified teachers and facilities (Weintrop et al., 2016). In conjunction with the marginalization of CS courses, integration of CT pedagogy into other subjects, specifically STEM subjects such as math and science, is uncommon (Grover & Pea, 2013). Nevertheless, a growing interest in this topic has led to some promising research on students' learning processes and outcomes from a CT pedagogy.

Research Findings on CT Pedagogy

Using the CT approach in math and science classrooms is necessary given a growing demand for computing skills in STEM fields in the digital age (Weintrop et al., 2016). A literature review by Weintrop et al. (2016) documented that math and science classrooms with CT pedagogy enabled students to access a more visualized view of math and science concepts and thus helped students grasp difficult or complex concepts more easily than in traditional math and science classrooms. In fact, previous studies showed that physics and biology courses incorporating CT helped students see scientific phenomena more intuitively while gaining a more organized understanding of scientific and mathematical concepts (Sengupta, Kinnebrew, Basu, Biswas, & Clark, 2013). Similarly, the literature showed that students in algebra classrooms with computational tools such as visualization (animation) can construct their own algebra knowledge by creating certain forms of data representation (Sengupta et al., 2013).

In an effort to prepare school-age students for a computational environment, stakeholders in the CS community have developed a wide range of user-friendly computer programming languages for novice learners, such as Scratch and Alice, which entail problem-solving and scientific inquiry (Weintrop et al., 2016). Research suggests that such tools are beneficial for young students in the development of CT and familiarity with a computational environment. For example, urban youth from underserved backgrounds were excited to play with Scratch, which was introduced in 2005, where they learned programming concepts, such as user interface, loops, and conditional statements (Maloney, Peppler, Kafai, Resnick, & Rusk, 2008). Through Scratch, urban youth in an after-school program used various programming activities to create animated stories, build roller coaster games, and manipulate images in Adobe Photoshop. The Maloney et al. (2008) findings were derived from 536 Scratch programs created by urban youth aged 8 through 18 years in low-income communities with limited access to computers. Another similar computer programming language for the novice learner is Alice, which is operated in a three dimensional interactive animation environment (Cooper, Dann, & Pausch, 2000). Given the traditional lack of exposure to computers and computing culture for girls, Kelleher, Pausch, and Kiesler (2007) investigated how Alice influenced the middle school girls' learning. In this study, a total of 88 girls participating in both storytelling and generic Alice. Researchers found that both types of Alice equally engaged girls in learning programming concepts, while those with storytelling Alice were more likely to spend additional time programming and show greater interest in using Alice in the future compared with their generic Alice counterparts.

A number of studies investigated such student cognitive skills as reasoning, logic, and problem-solving as well as motivation, engagement, and interest in CT pedagogy. For example, a meta-analysis of 65 studies showed that students who were exposed to computer programming languages showed 16% higher scores on cognitive-ability tests compared to their counterparts (Liao & Bright, 1991). Bers, Flannery, Kazakoff, and Sullivan (2014) found that the TangibleK Robotics Program, which is embedded with computer programming and robotics, improved kindergarten children's interest in various activities related to CT including debugging, correspondence, sequencing, and control flow. In a study by Israel, Pearson, Tapia, Wherfel, and Reese (2015), teachers reported that struggling students, such as students from low income backgrounds and students with disabilities, collaborated and encouraged each other to engage in solving challenging problems through various computing activities.

However, to date, relatively few studies have investigated the direct effects of such pedagogy on student learning outcomes related to college and career readiness. One study on STEM major selection (Lee, 2015a), after taking into account several well-documented predictors for STEM major selection, such as American College Testing (ACT) scores on math and science, found that students taking more credits for computer sciences were more likely to choose STEM majors. Another study provided evidence that students who frequently engaged in video or computer-game based activities were more likely to choose STEM majors compared with their counterparts (Lee, 2015b). Video or computer-game based activities demand problem-solving activities through a multidimensional screen, which share common characteristics with user-friendly computer programming languages like Scratch and Alice.

Practical Challenges for Implementing CT Pedagogy

Several barriers block the implementation of CT pedagogy in K–12 classrooms. First is a widespread misconception of CT in the general public, which undermines building facilities and training qualified teachers to promote such a reformed curriculum. Other barriers are rooted in limited financial resources for reforming the current form of K–12 curriculum (Topper & Lancaster, 2013). Specifically, schools serving low-income students typically are grappling with budget shortfalls and a high teacher turnover rate, long-standing problems that do not allow stakeholders to initiate curriculum reform.

A second major concern is the marginalization of CS courses, which are foundational to incorporating CT pedagogy in K–12 classrooms. In the ongoing discussion of how to generalize CS education at the secondary level, 20 states allow CS credits to count toward a math or science course (Loewus, 2016). However, the National Council of Teachers of Mathematics raised a concern that this allowance could leave students underprepared for math proficiency necessary for college and career readiness (Loewus, 2016). Diane Biars, President of the National Council of Teachers of Mathematics, claimed that CS courses should be added as independent core courses for STEM education, rather than count as math or science courses (Loewus, 2016). However, treating CS as an independent or elective course often leads to schedule constraints (K–12 Computer Science Framework, 2016). A strong consensus has yet to be established for incorporating CT pedagogy into K–12 classrooms.

MUTUAL BENEFITS FROM PARTNERSHIPS BETWEEN K–12 SCHOOLS, UNIVERSITIES, AND COMMUNITIES

A strong consensus on reforming the K–12 curriculum toward CT pedagogy could be established when each stakeholder is aware of its own benefit from the reform-oriented curriculum. Research partnerships among K–12 stakeholders, which focus on student learning processes and outcomes, could be an initial step toward curriculum reform. This section discusses potential benefits of universities, K–12 schools, and communities through building research partnership.

Potential Benefits of Universities, K–12 Schools, and Communities

Building research partnerships, such as R&D agreements and research consortia, is a recommended approach to provide resources and reduce costs for higher education institutions (Hillman, Withers, & Collins, 2009). Given the projection that higher education faces revenue short-falls (Moody's Investors Service, 2017), there is a growing demand for external funding. For example, public research universities grapple with the continual decline of state appropriations (National Science Board, 2012). Tuition-driven colleges, such as small liberal arts colleges, also face financial challenges due to significant drops in enrollment (Marcus, 2017). In fact, college enrollments across the nation declined continuously from 2012 to 2017, which is detrimental to all institutions of higher education (National Student Clearinghouse Research Center, 2017). As such, to increase enrollments and revenue, colleges and universities are demanding more publications and research grants from faculty, which positively influences institutional rankings (Davis, 2016). Accordingly, their best interests are in developing promising research agendas. A research agenda for incorporating CT pedagogy in K–12 classrooms could be in the best interests of numerous funding organizations and employers that demand a technologically literate workforce. From the K–12 stakeholders' standpoint, providing professional development for teachers in CT pedagogy as well as building the appropriate technological infrastructure are the immediate benefit of resources earned from research partnerships with universities, specifically in low-income communities. In the long term, computing education would promote economic growth through developing a skilled technical work-force and meeting the recruiting needs of employers.

Cases of Research Partnerships in Incorporating CT Pedagogy Into K–12 Classroom

As framed by the above snapshot of the potential benefits of research partnerships, this section describes ten cases of research partnerships among K–12, universities, and communities, funded by the National Science Foundation (NSF), which designed reforms for CT pedagogy in K–12 class-rooms. Please note that among the awarded research projects, this study pulled the abstracts for research projects that expired prior to January 1,

2016 to locate certain forms of written documents that indicated research processes or outcomes. Among U.S.-based, awarded NSF research projects in the *Directorate for Education and Human Resources* (National Science Foundation, 2018), awarded from 2000 to 2015, ten cases pertaining to CT pedagogy in K–12 classrooms were randomly selected and examined for the benefits of the research partnerships to diverse stakeholders. The 10 awarded cases are listed and analyzed in chronological order from earliest to latest. Table 11.1 presents a summary of the awarded cases in light of the grant period, stakeholders in partnership, and summary of each case's research project.

Table 11.1.
Summary of 10 Awarded Research Projects from 2000 to 2015

Case #	Grant Period	Stakeholders in the Partnership	Summary
Case 1	10/1/00 – 9/30/04	K–12 Schools (8th–12th grade teachers); Palm, Inc.; Texas Instrument	The project investigated the way in which math classrooms incorporating school-standard graphing calculators and interactive computational devices influenced 7th to 9th graders' math performance. A study shows that the students participating in this project made significant gains from pre/posttests drawn from a rigorous 10th grade state examination (Hegedus & Kaput, 2002).
Case 2	6/15/04 – 5/31/10	K–12 Schools; University of Wisconsin – Madison	This project explored the effects of computational tools, such as the epistemic game *Digital Zoo*, on high school students' science achievement scores and problem-solving skills. High school students participating in this project showed a significant gain in test scores on physics and biology questions from their textbook (up 600% on average) (Shaffer & Gee, 2005). Further, after completing another epistemic game *Madison 2200*, students showed a better problem-solving skill through gaining 70% more in the domain of complex thinking ability on urban planning (Shaffer & Gee, 2005).

(Table continues on next page)

**Table 11.1.
(Continued)**

Case #	Grant Period	Stakeholders in the Partnership	Summary
Case 3	9/15/06 – 8/31/12	The Chicago Public School; Beecher High School; Northwestern University; University of Chicago; University of Pennsylvania	This project developed *CogSketch* which is a visual-spatial sketch pad designed for enhancing reasoning and communication skills necessary for the 21st century workforce. The research team found that 5th graders gained a better understanding of the circulatory system when comparing their knowledge between before and after using the *CogSketch* (Miller et al., 2013). Beyond this preliminary finding, in a review of literature by Newcombe (2013), several studies provided evidence that learning materials integrating into a visual-spatial computational tool like *CogSketch* contributed to the improvement of K–12 students' visual spatial abilities, their performance on STEM-related tasks, and their likelihood of pursuing STEM careers.
Case 4	9/15/06 – 8/31/13	A-C Central Community Unit School District #262, and Regional Office of Education (ROE) #38; Rural ROEs (55, 3, 25, & 11); The Illinois Institute for Rural Affairs; Illinois Petroleum Resources Board; Illinois Science Teachers Association;	This research project provided high school chemistry teachers with professional development (PD) for integrating computational tools and contents into their traditional classrooms. The research team investigated the effects of the computational-based chemistry curriculum on high school students' knowledge gain. A study shows that students who were taught by experimental group teachers participating in the PD gained a higher score on the American Chemical Society High School Chemistry Test compared to their control group counterparts (Murray et al., 2009).

(Table continues on next page)

**Table 11.1.
(Continued)**

Case #	Grant Period	Stakeholders in the Partnership	Summary
Case 4		National Center for Rural Health Professions; Three Rivers Educational Partnership; The National Board Resource Center at Illinois State University; Argonne National Laboratory;	
		University of Washington – Seattle;	
		University of Illinois at Urbana-Champaign;	
		AC Central School District; National Center for Supercomputing Applications; MSPnet	
Case 5	9/15/08 – 6/30/13	Phoenix Union High School District; Scottsdale Union High School District; Roosevelt District;	This project was designed to promote computational thinking abilities of adolescent girls of color from underserved communities through providing a culturally relevant technology program. A study found that the underrepresented girls of color, who had little or no access to computers prior to participating in this project, were highly engaged in multimedia activities to analyze such social issues as problems related to alcoholism and early motherhood; further, the girls of color became interested in becoming computing professionals after engaging in various hands-on and technology-based learning activities (Scott & White, 2013).

(Table continues on next page)

Table 11.1.
(Continued)

Case #	Grant Period	Stakeholders in the Partnership	Summary
Case 5		Boys and Girls Club; of the East Valley-Sacaton;	
		Intel; Applied Learning Technologies Institute;	
		Dynamic Educational Leadership for Teachers and Administrators (D.E.L.T.A.); Arizona State University (ASU)'s School of Computing & Informatics; ASU's Video Game Design Camp; Arizona Council of Black Engineers and Scientists Computer Camp (ACBES);	
		Carnegie Mellon University	

(Table continues on next page)

**Table 11.1.
(Continued)**

Case #	Grant Period	Stakeholders in the Partnership	Summary
Case 6	5/1/09 – 8/31/15	Local Two High Schools (Gallagher et al., 2011); University of Colorado at Boulder	In this project, the research team developed a computing-based curriculum and incorporated such a reform-based curriculum into traditional middle and high school classrooms. A study showed mixed findings on the effects of an advanced biology curriculum embedded with computational algorithms as follows: (1) the student participants were able to understand, analyze, and interpret existing algorithms after completing the program, while struggling to apply their existing algorithm knowledge to solving other problems; and (2) some students were excited to learn more about computational contents to prepare for a career in biology, while others raised a question of why they had to be engaged in such a computational algorithms in advanced biology, especially when encountering academic challenges to apply their existing algorithm knowledge to solving other problems (Gallagher et al., 2011).
Case 7	9/1/09 – 8/31/15	Middle School; New Mexico State University	The research team developed computer mediated animations and games for the Math Snack Program, which aims at enhancing conceptual understanding in middle school level math and developing mathematical thinking and communication skills. In a study by Valdez et al. (2013), 7th graders assigned to teachers who implemented the Math Snack Program showed a significant gain on pre/posttest as measured by math test-items from various standardized tests, including the National Assessment of Education Progress (NAEP); however, there was no significant difference in math score gain among 6th graders.

(Table continues on next page)

Table 11.1.
(Continued)

Case #	Grant Period	Stakeholders in the Partnership	Summary
Case 8	11/1/09 – 7/31/11	St. Joseph's University in Philadelphia, PA; Ithaca College; Santa Clara University; Duke University; Colorado School of Mines; Virginia Beach School System; Association of Computing Machinery	The research team provided PD for teachers to use a computer software program called Alice in a wide range of classrooms including math, science, history/social studies, and English. Through using Alice, students are involved in creating computer graphics, animation, and storytelling. Preliminary findings included that (1) middle school students' enrollment in the Alice course continued its steady rise over a three-year period from 2005 to 2008, especially with girls and minority students; (2) 4th to 6th grade girls were excited to play with Alice, and (3) teachers and parents perceived that through engaging in Alice-based learning in various classrooms, students were motivated to learn computer programing as well as topics covered in the various classrooms (Rodger et al., 2009).
Case 9	9/1/10 – 8/31/14	University of New Mexico; New Mexico Tech; New Mexico State University; Massachusetts Institute of Technology (MIT); Santa Fe Complex; Girl Scouts of New Mexico Trails; Supercomputing Challenge; Regional educational organizations, and local schools	In an effort to promote computer science education for an underserved student group, primarily Hispanic/Latino and low-income female students, the research team asked students to use a user-friendly computer programming language program, such as Scratch. A preliminary study described students' learning processes in light of engagement in the computer programming languages; students built a digital-story in language arts and history classes using Scratch, students investigated computational science problems in the science classroom using StarLogo, and students collected and analyzed data using iSENSE (Lee et al., 2014).

(Table continues on next page)

**Table 11.1.
(Continued)**

Case #	Grant Period	Stakeholders in the Partnership	Summary
Case 10	7/1/11 – 6/30/15	Four urban schools in the northeast region of the U.S (Bodzin et al., 2014); Lehigh University	This project investigated the way in which integrating Web Geospatial Information Systems (GIS) into an earth science classroom influence middle school students' STEM spatial thinking skills and understanding of earth science contents. A study found that the GIS-based earth science curriculum significantly improved students' tectonics content knowledge as well as the geospatial thinking and reasoning skills from pre/posttests with a large effect size (Bodzin et al., 2014). In another study, teacher participants perceived the GIS-based curriculum contributed to the improvement of their pedagogical content knowledge as well as students' understanding on earth science contents (Bodzin, 2015).

Case 1: Understanding Classroom Interactions Among Diverse, Connected Classroom Technologies (Grant # REC-0087771)

The grant period for this research project was October 1, 2000 to September 30, 2004. A research team from the University of Massachusetts Dartmouth and SRI International collaborated with teachers in grades 8 through 12 who taught in math classrooms equipped with school-standard graphing calculators and interactive computational devices wirelessly networked to each other and to a teacher's workstation. Corporate partners included Palm, Inc., which designed the wirelessly connected classrooms and Texas Instruments, which developed the *TI-83* Plus graphing calculator (Kaput & Schorr, 2008). This project sought to explore student learning processes–encompassing students' responses, contributions to the classroom, and the teacher-student relationship—as well as how teachers managed the classroom and supported student learning.

The research team published several articles in books, conference proceedings, and peer-reviewed journals. The articles addressed the ways students and teachers benefit from the wirelessly networked devices in math classrooms. For example, Hegedus and Kaput's (2002) study provided evidence that the wirelessly connected classroom equipped with hand-held devices, such as the *TI-83* Plus graphing calculators, enabled seventh to ninth grade students ($N = 35$) to engage in small-group discussion through sharing mathematical objectives, such as slope-as-rate, linear functions, and simultaneous equations. The student engagement through active learning in basic algebra resulted in significant gains from pre/posttests drawn from a rigorous state 10th grade examination, especially with low-performing students. Further, through this research partnership, the corporate partners, Palm, Inc. and Texas Instruments, immediately gained the opportunity to advertise their products and services.

Case 2: CAREER, Alternate Routes to Technology and Science (GRANT # REC-0347000)

The grant period for this research project was May 27, 2004 to May 31, 2010. The principal investigator (PI) recruited graduate research assistants and provided them with training and education for conducting research, which was part of the PI's education plan in an NSF Faculty Career Award program.

The research team explored how high school students gain an understanding of fundamental STEM concepts and skills using computational tools, such as the epistemic game *Digital Zoo* (Svarovsky & Shaffer, 2007). The research resulted in several publications including conference proceedings, peer-reviewed journal articles, and a book. For example, student participants role-played as biomechanical engineers through designing virtual structures and creatures in the game (Shaffer et al., 2009). The research team found that students learned physics and biology beyond learning the tasks of biomechanical engineers as evidenced by a significant gain in test scores (up 600% on average) on science questions from their textbook after playing the game (Shaffer & Gee, 2005). Similar to the concept of the *Digital Zoo*, the investigators developed a new epistemic game called *Madison 2200*, which was a computer-based game for players to learn urban ecology through their involvement in various roles and responsibilities as urban planners, such as redesigning a downtown pedestrian mall, planning a city budget, and assessing community needs (Shaffer, Halverson, Squire, & Gee, 2005). The research team found that after playing *Madison 2200*, eleven high school students gained 70% in complex thinking scores and considered 20% more factors that could significantly affect urban planning, suggesting that the students had better comprehension for solving real-world complex problems pertaining to the intersection of science, society, economics, and technology (Shaffer & Gee, 2005).

Case 3: Spatial Intelligence and Learning Center (GRANT # SBE–0541957)

The grant period for this research project was September 15, 2006 to August 31, 2012. The research team constituted researchers from multiple disciplines, including cognitive science, psychology, computer science, education, and neuroscience and practicing geoscientists and engineers. The interdisciplinary research team, in collaboration with the Chicago Public Schools, sought to provide depth of understanding on spatial learning processes and outcomes for children and adolescents and develop programs and technologies that could advance a wide range of reasoning and communication skills necessary for 21st century citizens (including but not limited to problem solving skills in math, spatial abstraction and design, and using spatial metaphor in mental models of complex domains). Further, the research team produced the following learning materials: (1) a new sketch-based learning tool called *CogSketch* that can support reform-oriented pedagogies, (2) a new assessment battery for assessing spatial skills, and (3) a curriculum embedded with spatial contents related to geoscience and engineering. Moreover, this research project provided cross-disciplinary training opportunities for various groups, such as junior scientists and teachers.

The research team found that fifth grade students ($N = 49$) gained a significantly better understanding of the circulatory system after using the *CogSketch*, which is a visual-spatial sketch pad (Miller, Cromley, & Newcombe, 2013). In this study, the students were asked to label and draw arrows showing the movement of blood, oxygen, and carbon dioxide on the *CogSketch* worksheets and received feedback from *CogSketch* whenever they requested it. Further, Newcombe (2013) documented evidence from several studies showing that learning materials incorporating a visual-spatial computational tool like *CogSketch* increased K–12 students' visual spatial abilities, showed better performance on STEM-related tasks, and increased students' likelihood to pursue STEM careers.

Case 4: Institute for Chemistry Literacy Through Computational Science (ICLCS) (Grant # 0634423)

The grant period for this research project was September 15, 2006 to August 31, 2013. The research team comprised multiple stakeholders, including the Department of Chemistry, the National Center for Supercomputing Applications, and the College of Medicine at the University of Illinois–Urbana Champaign and six school districts located in the northwestern, west-central, south-central, and far south regions of

Illinois. The overarching goals of the project included (1) to provide a better understanding of chemistry pertaining to scientific research; (2) to increase teachers' use of computational and visualization teaching tools; (3) to develop a community for advocating excellence in science education constituting 9th to 12th grade teachers and university-level faculty; and (4) to promote university-school district partnerships for reforming science classrooms necessary for 21st century learners. Using a randomized control trial, the research team explored the teaching process and chemistry learning through the outreach program. They published their findings in a peer-reviewed journal and their workshop summary in an edited book (Honey, Pearson, & Schweingruber, 2014).

The researchers conducted a quasi-experimental study to assess the effects of a computational-based curriculum in chemistry on high school students' knowledge (Murray, Henry, & Hogrebe, 2009). For this study, the research team provided professional development (PD) for chemistry teachers to integrate computational tools and content into their existing classroom. Students ($N = 963$) who were taught by the experimental teachers ($N = 29$) showed significant gains on the pre-/posttest of the American Chemical Society High School Chemistry Test in comparison with the control group of teachers ($N = 25$) and students ($N = 861$).

Case 5: COMPUGIRLS, a Culturally Relevant Technology Program for Girls (Grant # EHR–0833773)

The grant period for this research project was September 15, 2008 to June 30, 2013. Members from the following institutions formed the research team: Arizona State University (ASU), ASU's School of Computing & Informatics, ASU's Video Game Design Camp, Phoenix Union High School District, Roosevelt District, Scottsdale Union High School District, Boys and Girls Club of the East Valley-Sacaton, Intel, Applied Learning Technologies Institute, Dynamic Educational Leadership for Teachers and Administrators (D.E.L.T.A.), and the Arizona Council of Black Engineers and Scientists Computer Camp (ACBES). The project provided underserved adolescent girls (grades 8 to 12) with hands-on, technology-based learning activities designed to develop computational thinking and encourage them to pursue ICT/STEM fields. Many of the girls had never used a computer prior to participating and came from marginalized backgrounds, including a group of students who dropped out of high school and/or teen mothers. Theoretical lenses included Culturally Relevant Pedagogical Practices (CRP), the Social Justice Youth Development Framework, and the Future Time Perspective. Specifically, student participants took six multimedia courses: (1) social justice and the media,

(2) Scratch, user-friendly computer programming, (3) modifying SIMS (designing, modifying, and troubleshooting simplified simulations), (4) the choice of 3-D, programming, and character design, (5) an advanced choice of 3-D, programming, and character design, and (6) teamwork to create neighborhoods in Sim City using all skills learned. Students used various computing programs and hardware technologies to analyze real-world, social problems such as the consequences of early motherhood or alcoholism, and then shared their findings through internships, conference and community presentations, and parent workshops.

The research team described the practical implications of COMPU-GIRLS, which is a culturally relevant technology program (Scott, Clark, Sheridan, Mruczek, & Hayes, 2010). They also offered three guidelines from culturally responsive pedagogical theory: (1) reflective action: teachers must show a respectful attitude toward students' of color cultural contents which are permeable to the classroom; (2) asset building approach: teachers should identify educational needs and interests tailored to the cultural backgrounds of students of color; and (3) connectedness: teachers should cultivate a sense of connectedness between students and their communities. A second publication documented that African-American and Latino female students ($N = 41$) were highly motivated to pursue computer-related careers after taking the six courses (Scott & White, 2013). As such, the finding subverted the myth that girls of color are just not interested in engaging in computing activities and pursuing STEM careers in general.

Case 6: New, GK-12, Integrating Computer Science into Traditional Studies (GRANT # DGE–0841423)

The grant period for this project was May 1, 2009 to August 31, 2015. Partnering with local school districts and public school teachers, the project team intended to incorporate computing into traditional middle and high school courses (such as biology, physics, civics), in-school technical electives, summer programs, and research projects at the high school level. Working collaboratively, the partners developed computing-based curriculum tailored to the needs of each individual school. The project's overarching goals were to promote computing skills and knowledge to a broader population with the following steps: (1) cultivate computing education in both K–12 and higher education settings; (2) advance understanding of the concepts and applications of computing to the public; and (3) eventually increase the number of computer professionals in the workforce.

The research team published a study that explored how students perceived and learned from a high school, advanced biology curriculum that was integrated with computational algorithms, such as the Basic Local

Alignment Search Tool (BLAST) (Gallagher et al., 2011). Scholars found that students could understand, analyze, and interpret existing algorithms after completing the program, but were unable to apply their existing algorithmic knowledge to solve other cases. Gallagher et al. (2011) found that students enjoyed taking the first module that demonstrated how computational algorithms are used to solve biology-related problems; further, teachers reported that some students showed interest in learning more about computational contents, perceiving that these are necessary to becoming a real biologist. However, when students encountered challenges in solving their own problems, they asked why they had to learn biology incorporating computational algorithms; further, several students were concerned that such curriculum was not directly tied to the questions on the AP or IB tests. As such, the research team pointed out that a computational-based curriculum should be incorporated into the AP or IB tests.

Case 7: Math Snacks, Addressing Gaps in Conceptual Mathematics Understanding With Innovative Media (GRANT # 0918794)

The grant period for this project was September 1, 2009 and August 31, 2015. In collaboration with public middle schools in New Mexico, the research team led by math and education faculty at the New Mexico State University developed computer-mediated animations and games for the Math Snack program. They designed the modifications to advance conceptual understanding of middle grades mathematics and to enable middle school students to develop mathematical thinking and communication skills. A pilot study in Year 3 preceded a randomized control trial in Year 4 to assess the effects of using the Math Snack software on student math learning outcomes as measured by math scores on New Mexico state mathematics assessments.

In the pilot study, the research team investigated the effects of the Math Snack program on middle school students' math performance on test items from various standardized tests, such as National Assessment of Education Progress (NAEP) (Valdez, Trujillo, & Wiburg, 2013). Findings from this study concluded that the experimental group of seventh grade students taught by teachers using the Math Snack guideline ($N = 58$) showed significant pre/posttest gains in math performance compared with the control group ($N = 67$), while no significant differences emerged between the experimental ($N = 189$) and control groups ($N = 146$) for sixth grade students. Beyond the study findings, the research team documented how they built research partnerships between university and public schools to improve math learning environments for students (Kinzer, Wiburg, &

Virag, 2010). Moreover, they provided a practical guide of how game-based instruction like the Math Snack program can be effectively facilitated in K–12 classrooms (Chamberlin, Trespalacious, & Gallagher, 2012).

Case 8: An Innovative Approach for Attracting Students to Computing: A Comprehensive Proposal (NSF #ESI– 0624642, 0624654, 0624528, & NSF supplement DRL– 0826661)

This research project began on November 1, 2009 and ended on July 31, 2011. Stakeholders in this three-year project included five higher education institutions, one school system, and the Association of Computing Machinery (ACM). The project's overarching goal was to engender K–12 students' interest in learning computer science, specifically targeting girls and underrepresented minorities. To achieve this goal, the research team asked high school students to use a software program called Alice, which is operated by 3-D visualization. Through their involvement in creating computer graphics, animation, and storytelling, the high school students learned how to perform object-based programming. The high school provided PD for teachers to incorporate Alice in their teaching practices and built a network with college faculty to offer continuing support for the teachers during the academic year.

The research team published the findings about this teaching and learning phenomenon using Alice in conference proceedings. For example, the research team provided a practical guide of how to use the Alice in K-12 classrooms and where to download the learning materials (see the following link: https://www2.cs.duke.edu/csed/alice/) and software (see the following link: http://www.alice.org/get-alice/) for free (Rodger et al., 2009). Rodger and colleagues (2009) also presented their preliminary findings about student and teacher engagement using Alice in a diverse set of science, math, history/social studies, and English courses. The findings included: (1) the gradual increase from 2005 to 2008 in the number of middle school students enrolling in the Alice courses, specifically girls and minority students; (2) fourth to sixth grade girls expressed excitement after using Alice; and (3) teachers and parents perceived Alice as a useful computational tool for young students to learn computer programming in a fun and exciting way. Rodger and colleagues noted that the Alice courses covered a range of computer science topics, including parameters, loops, if statements, and other basic topics for technology literacy, such as vehicle property, camera controls, and color property.

Case 9: Strategies, GUTS y Girls (GRANT # ITEST–1031421)

This research project began on September 1, 2010 and ended on August 31, 2014. The project emerged from a previously funded NSF Academies for Young Scientist award (06-39637). Numerous stakeholders partnered in this project, including universities, K–12 schools, industry, and non-profit organizations associated with computer science education. The project aimed to broaden participation in the nation's STEM and ICT workforce by motivating underserved female students in grades 5–8 to learn CT. The design promoted computer science education for primarily Hispanic/ Latino, low-income, female students through hands-on design-and-build activities in a user-friendly computer programming language program, such as Scratch. Further, the research team developed a social network support system for female students by creating a virtual clubhouse where the female students could communicate with female STEM/ICT practitioners and student mentors. Leaders tailored the research activities to the New Mexico state standards in science and math for grades.

The research team published the results of their project on a dedicated website (projectguts.org) (Lee, Martin, & Apone, 2014). The published article described three examples of using computational activities in the domains of building a digital story, applying preexisting computer models to create new computer models to solve real-world problems, and exploring data from data generation to analysis. The three examples presented middle school students' work using a user-friendly computer programming language as follows: (1) using Scratch for digital storytelling in language arts and history classes; (2) using StarLogo for computational science investigation in science classroom; and (3) using iSENSE for data collection and analysis. While the published article did not provide student learning outcomes, it detailed how students learned computational thinking abilities through engaging in a series of computational activities.

Case 10: Promoting Spatial Thinking With Web-Based Geospatial Technologies (GRANT #1118677)

This partnership of universities, middle school teachers, and instructional designers addressed a research project that began on July 1, 2011 and ended on June 30, 2015. The project design aimed to improve middle school students' STEM spatial thinking skills through incorporating Web Geospatial Information Systems (GIS) into earth science classrooms. GIS constitutes advanced visualization and geospatial analysis functions. As an overarching goal, the project investigated how Web GIS could serve as best practices for enhancing both spatial thinking skills and earth science learning outcomes for diverse learners. To achieve the goal, the research team examined how teachers implemented the earth science curriculum equipped with GIS and assessed its effects on student learning outcomes.

The research team documented that students with the GIS-based curriculum showed a significant gain on tectonics content knowledge and geospatial thinking and reasoning from pretest to posttest with large effect sizes in a sample of 1,124 students (Bodzin, Fu, Bressler, & Vallera, 2014). Further, in survey responses and focus group interviews, the 12 eighth grade teachers involved perceived that the Web GIS-based curriculum enhanced their pedagogical content knowledge and students' understanding of earth science concepts and processes (Bodzin, 2015).

A Conceptual Model of Mutual Benefits From Research Partnerships

The reviewed 10 cases reflect the common benefits for universities, K–12 schools, and communities, regardless of each project's purpose and design, which suggests a conceptual model of mutual benefits from research partnerships (Figure 11.1). From the university side, a partnership can fulfill higher education's mission of contributing to local and national communities through helping students develop the CT skills necessary for tomorrow's workforce. Further, universities can advance knowledge through research activities. Specifically, partnerships can yield research grants and publications, which are directly beneficial for promoting university rankings and thus, increasing student enrollment, regardless of the type of post-secondary institution. From the K–12 schools' standpoint, as described by the various publication forms, research grants can support teachers' PD to advance their content and pedagogical knowledge pertaining to CT. Moreover, students' exposure to the CT-based learning environment can serve as an antecedent for more students to become familiar with and/or interested in learning CT-related tasks, which could help students prepare for today's workforce. From the community's standpoint, IT-related corporations or experts are typically involved in developing CT-based learning tools or infrastructure, which generate their revenue sources. More broadly, a wide range of community members, including IT-related corporations and other employers, are building a STEM or computing workforce, which is projected to be continuously in demand.

It is hoped that this chapter motivates universities and communities to build research partnerships with K–12 schools to promote CT-related learning for all students. Such a partnership can enhance CT skills for today's students, which can strengthen tomorrow's workforce and ultimately, contribute to national economic growth. Such a well-prepared workforce can

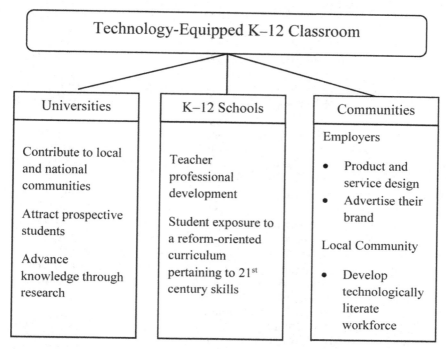

Figure 11.1.   A conceptual model of mutual benefits from research partnerships.

meet the growing demand of solving complex real world problems as listed previously, such as prevention or cure of diseases and reduction of world hunger. As such, key stakeholders who can be influential to integrate a CT pedagogy into K–12 classrooms need to initiate and develop partnerships for popularizing the idea of adopting a CT pedagogy in K–12 classrooms.

REFERENCES

Barr, V., & Stephenson, C. (2011). Bringing computational thinking to K–12: What is involved and what is the role of the computer science education community? *ACM Inroads, 2*(1). 48–54. doi.org/10.1145/1929887.1929905

Bers, M.U., Flannery, L.P., Kazakoff, E.R., & Sullivan, A. (2014). Computational thinking and tinkering: Exploration of an early childhood robotics curriculum. *Computers & Education, 72*, 145-157. doi.org/10.1016/j.compedu.2013.10.020

Bodzin, A. M. (2015, March). *Using educative curriculum materials to support teacher enactment of a geospatial science curriculum.* Paper presented at the annual meeting of the Society for Information Technology and Teacher Education

(SITE), Las Vegas, NV. Retrieved from http://www.ei.lehigh.edu/eli/research/site2015.pdf

Bodzin, A. M., Fu, Q., Bressler, D., & Vallera, F. L. (2014, March). *Examining the enactment of Web GIS on students' geospatial thinking and reasoning and tectonics understandings.* Paper presented at the Annual International Conference of the National Association for Research in Science in Science Teaching, Pittsburgh, PA. Retrieved from http://www.ei.lehigh.edu/eli/research/narst2014.pdf

Chamberlin, B., Trespalacios, J. S., & Gallagher, R. (2012). The Learning Games design model: Immersion, collaboration, and outcomes-driven development. *International Journal of Game-Based Learning (IJGBL)*, 2(3), 87–110. doi.org/10.4018/ijgbl.2012070106

Cooper, S., Dann, W., & Pausch, R. (2000). Alice: A 3-D tool for introductory programming concepts. *Journal of Computing Sciences in Colleges, 15*(5), 107-116.

Davis, M. (2016). Can college rankings be believed? *She Ji: The Journal of Design, Economics, and Innovation, 2*(3), 215–230. doi.org/10.1016/j.sheji.2016.11.002

Gallagher, S. R., Coon, W., Donley, K., Scott, A., & Goldberg, D. S. (2011). A first attempt to bring computational biology into advanced high school biology classrooms. *PLoS Computational Biology*, 7(10), e1002244. http://doi.org/10.1371/journal.pcbi.1002244

Grover, S., & Pea, R. (2013). Computational thinking in K–12: A review of the state of the field. *Educational Researcher, 42*(38). 38–43. doi.org/10.3102/0013189X12463051

Hegedus, S., & Kaput, J. J. (2002, October). Exploring the phenomenon of classroom connectivity. *Proceedings of the 24th Annual Meeting of the North American Chapter of the International Group for the Psychology of Mathematics Education*, Athens, GA. Retrieved from http://www.kaputcenter.umassd.edu/downloads/simcalc/cc1/library/jksh_pmena2002.pdf

Hillman, A. J., Withers, M. C., & Collins, B. J. (2009). Resource dependence theory: A review. *Journal of Management, 35*(6). 1404–1427. doi.org/10.1177/0149206309343469

Hoffmann, L. (October, 2012). CS and the three Rs. *Communications of the ACM, 55*(1), 17–19.

Honey, M., Pearson, G., & Schweingruber, H. (Eds.). (2014). *STEM integration in K-12 education: Status, prospects, and an agenda for research.* Washington, DC: National Academies Press.

Israel, M., Pearson, J. N., Tapia, T., Wherfel, Q. M., & Reese, G. (2015). Supporting all learners in school-wide computational thinking: A cross-case qualitative analysis. *Computers & Education, 82*, 263–279. doi.org/10.1016/j.compedu.2014.11.022

K–12 Computer Science Framework. (2016). Retrieved from http://www.K12cs.org.

Kaput, J., & Schorr, R. (2008). Changing representational infrastructures changes most everything. The case of SimCalc, algebra, and calculus. *Research on technology and the teaching and learning of mathematics. Cases and perspectives* (Vol. 2, pp. 211–253). Charlotte, NC: NCTM and Information Age.

Kelleher, C., Pausch, R., & Kiesler, S. (2007, April). Storytelling Alice motivates middle school girls to learn computer programming. *Proceedings of the SIGCHI conference on Human factors in computing systems* (pp. 1455–1464). ACM.

Kinzer, C., Wiburg, K., & Virag, L. (2010). University public school research partnerships in mathematics. *Border Walking Journal, 8*(1), 61–69.

Lee, A. (2015a). Determining the effects of computer science education at the secondary level on STEM major choices in postsecondary institutions in the United States. *Computers & Education, 88*, 241–255. doi.org/10.1016/j.compedu.2015.04.019

Lee, A. (2015b). An investigation of the linkage between technology-based activities and STEM major selection in 4-year postsecondary institutions in the United States: Multilevel structural equation modeling. *Educational Research and Evaluation, 21*(5), 439–465. doi.org/10.1080/13803611.2015.1093949

Lee, I., Martin, F., & Apone, K. (2014). Integrating computational thinking across the K-8 curriculum. *ACM Inroads, 5*(4), 64–71. doi.org/10.1145/2684721.2684736

Liao, Y. K. C., & Bright, G. W. (1991). Effects of computer programming on cognitive outcomes: A meta-analysis. *Journal of Educational Computing Research, 7*(3), 251–266. doi.org/10.2190/E53G-HH8K-AJRR-K69M

Loewus, L. H. (2016, September 13). More states mandate high schools count computer science as math or science. *Education Week*. Retrieved from http://blogs.edweek.org/edweek/curriculum/2016/09/computer_science_high_school_graduation.html

Maloney, J. H., Peppler, K., Kafai, Y., Resnick, M., & Rusk, N. (2008). *Programming by choice: urban youth learning programming with scratch* (Vol. 40, No. 1, pp. 367–371). ACM. doi.org/10.1145/1352322.1352260

Marcus, J. (2017, June 29). Many small colleges face big enrollment drops. Here's one survival strategy in Ohio. *The Washington Post*. Retrieved from https://www.washingtonpost.com/

Miller, B.W., Cromley, J.C., & Newcombe, N.S. (2013). Showcase May 2013: Exploration of CogSketch as an instructional tool in middle school science. *Spatial Intelligence and Learning Center*. Retrieved from http://spatiallearning.org/index.php/showcase/178-showcase-may-2013-exploration-of-cogsketch-as-an-instructional-tool-in-middle-school-science

Moody's Investors Service. (2017). *Higher Education–US: 2018 outlook changed to negative as revenue growth moderates*. Retrieved from https://www.insidehighered.com/sites/default/server_files/media/2018%20Outlook%20for%20Higher%20Education%20Changed%20to%20Negative.pdf

Murray, K. S., Henry, M. A., & Hogrebe, M. C. (2009). The impact of early positive results on a mathematics and science partnership: The experience of the Institute for Chemistry Literacy through Computational Science. *The Journal of Mathematics and Science, 11*, 95–108.

National Education Association. (n.d.). *Preparing 21st century students for a global society: An educator's guide to the "Four Cs."* Retrieved from http://www.nea.org/assets/docs/A-Guide-to-Four-Cs.pdf

National Science Board. (2012). *Diminishing funding and rising expectations: Trends and challenges for public research universities*. Retrieved from https://www.nsf.gov/nsb/publications/2012/nsb1245.pdf

National Science Foundation. (2018). *Search Awards*. Retrieved from https://www. nsf.gov/awardsearch/

National Student Clearinghouse Research Center. (2017). *Current term enrollment—Fall 2017*. Retrieved from https://nscresearchcenter.org/current-term-enrollment-estimates-fall-2017/

Newcombe, N. S. (2013). Seeing relationships: Using spatial thinking to teach science, mathematics, and social studies. *American Educator, 37*(1), 26. Retrieved from https://www.aft.org/sites/default/files/periodicals/Newcombe_0.pdf

Rodger, S. H., Hayes, J., Lezin, G., Qui, H., Nelson, D., & Tucker, R. (2009). Engaging middle school teachers and students with Alice in a diverse set of subjects. *Proceedings of the 40th ACM Technical Symposium on Computer Science Education*. New York, NY.

Svarovsky, G., & Shaffer, D. W. (2007). SodaConstructing knowledge through exploratoids. *Journal of Research in Science Teaching, 44*(1). 133–153. doi.org/10.1002/tea.20112

Scott, K., Clark, K., Sheridan, K. M., Mruczek, C., & Hayes, E. (2010, March). Culturally relevant computing programs: Two examples to inform teacher professional development. In *Society for Information Technology & Teacher Education International Conference* (pp. 1269–1277). Association for the Advancement of Computing in Education (AACE).

Scott, K. A., & White, M. A. (2013). COMPUGIRLS' standpoint: Culturally responsive computing and its effects on girls of color. *Urban Education, 48*(5), 657–681.

Sengupta, P., Kinnebrew, J. S., Basu, S., Biswas, G., & Clark, D. (2013). Integrating computational thinking with K-12 science education using agent-based computation: A theoretical framework. *Educational and Information Technologies, 18*(2), 351–380. doi.org/10.1007/s10639-012-9240-x

Shaffer, D. W., & Gee, J. P. (2005). Before every child is left behind: How epistemic games can solve the coming crisis in education. WCER Working Paper No. 2005-7. *Wisconsin Center for Education Research (NJ1)*. Retrieved from https://files.eric.ed.gov/fulltext/ED497010.pdf

Shaffer, D. W., Halverson, R., Squire, K. R., & Gee, J. P. (2005). Video games and the future of learning. WCER Working Paper No. 2005-4. *Wisconsin Center for Education Research (NJ1)*. Retrieved from https://files.eric.ed.gov/fulltext/ED497016.pdf

Shaffer, D. W., Hatfield, D., Svarovsky, G. N., Nash, P., Bagley, E., ... Mislevy, R. (2009). Epistemic network analysis: A prototype for 21st century assessment of learning. *International Journal of Learning and Media, 1*(2), 1–22. doi.org/10.1162/ijlm.2009.0013

Tech & Learning. (2018, May 20). Twenty states pass legislation to expand access to K–12 computer science. *Tech & Learning*. Retrieved from https://www.techlearning.com/ed-tech-ticker/twenty-states-pass-legislation-to-expand-access-to-k-12-computer-science

Thompson, K. D., Martinez, M. I., Clinton, C., & Díaz, G. (2017). Considering interest and action: Analyzing types of questions explored by researcher-practitioner partnerships. *Educational Researcher, 46*(8), 464–473. doi.org/10.3102/0013189X17733965

Topper, A., & Lancaster, S. (2013). Common challenges and experiences of school districts that are implementing one-to-one computing initiatives. *Computers in the Schools, 30*(4). 346–358. doi.org/10.1080/07380569.2013.844640

Tran, Y. (2018). Computational thinking equity in elementary classrooms: What third-grade students know and can do. *Journal of Educational Computing Research*. Retrieved from https://doi.org/10.1177/0735633117743918

U.S. Bureau of Labor Statistics. (2017). *Occupational outlook handbook: Computer and information technology occupations*. Retrieved from https://www.bls.gov/ooh/computer-and-information-technology/computer-and-information-research-scientists.htm

Valdez, A., Trujillo, K., & Wiburg, K. (2013). Math snacks: Using animations and games to fill the gaps in mathematics. *Journal of Curriculum and Teaching, 2*(2). 154–161. doi.org/10.5430/jct.v2n2p154

Wang, J., Hong, H., Ravitz, J., & Moghadam, S. H. (2016). Landscape of K-12 CS education in the U.S.: Perceptions, access, and barrier. *Proceedings of the 47th ACM Technical Symposium on Computing Science Education* (pp. 645–650). ACM.

Weintrop, D., Beheshti, E., Horn, M., Orton, K., Jona, K., Trouille, L, & Wilensky, U. (2016). Defining computational thinking for mathematics and sciences classrooms. *Journal of Science Education and Technology, 25*(1), 127–147. doi.org/10.1007/s10956-015-9581-5

Wilson, C., Sudol, L. A., Stephenson, C., & Stehlik, M. (2010). *Running on empty: The failure to teach K–12 computer science in the digital age*. New York, NY: The Association for Computing Machinery, The Computer Science Teachers Association.

Wing, J. M. (2006). Computational thinking. *Communications of the ACM, 49*(3), 33–35. doi.org/10.1145/1118178.1118215

ABOUT THE AUTHORS

Rachel F. Adler, PhD, is an Assistant Professor in the Department of Computer Science at Northeastern Illinois University. She received her PhD from the Graduate Center of the City University of New York, where she examined the performance effects of computer-based multitasking. Her research interests are in the areas of human-computer interaction and computer science education. She is currently the principal investigator on an NSF STEM+C grant on incorporating computational thinking and coding into a preservice curriculum for future elementary and middle school science and math teachers. As part of the grant, she leads an interdisciplinary team of faculty and students to create computational thinking modules which are embedded into science and math education classrooms. In addition, she designed and taught an introductory computer science course which provides education students with a basic foundation in computer science. Dr. Adler is also coinvestigator on a pilot project for an NIH U54 grant to develop a mobile application to improve the symptom burden and quality of life among Hispanic women with breast cancer. Dr. Adler has published papers in journals including *Education and Information Technologies, Computers in Human Behavior, ACM Transactions on Computer-Human Interaction, International Journal of Human Computer Studies, Interacting with Computers, Contemporary Clinical Trials, Translational Behavioral Medicine, and Design for Health.*

Junghyun Ahn is a doctoral candidate in Instructional Technology and Media at Teachers College and a researcher on the after-school project at Teachers College Community School. She received an EdM in Mind, Brain and Education from Harvard Graduate School of Education and an MS in Applied Statistics from Teachers College. She also studied special education and is currently working as a UX researcher and content developer at a special education startup. Her research focuses on the use of embodied instruction to teach programming concepts (computational thinking) and technology integration into education for visually impaired students.

Whitney Bortz earned her BA in Humanities from Pepperdine University and began her career in education as a math teacher in Los Angeles, California. While still teaching, she began working part-time at a research center at UCLA. She developed a passion for research in education and, as a result, pursued a PhD in Education at Queen's University in Northern Ireland. Her dissertation research, situated in both U.K. and U.S. classrooms, focused on the relationship between teachers' theories of learning and their practices of assessment. Bortz worked at Radford University for four years as the Director of Assessment in the College of Education. In addition to supporting faculty in their assessment and accreditation efforts, she served as the principal investigator for two grant-funded projects that focused on teacher education and teacher evaluation. Currently, as Postdoctoral Research Fellow at Virginia Tech she works on two NSF-funded projects. Alongside colleagues in the Department of Computer Science, her research investigates the integration of computational thinking into middle school chemistry instruction. She also serves on a team in the Department of Engineering Education that is looking at funding models in STEM graduate programs and related experiences, and career outcomes of graduate students.

Susan Brown is currently the Interim Associate Dean of Research for the NMSU College of Education as well as the Director of the STEM Outreach Center. She is the principal investigator/coinvestigator of two NSF grants, a 21st CCLC grant, several foundation grants, and various state/federal grants. Besides her grant work, she has taught science and math methods classes at New Mexico State University, as well as classes in the art of grant writing. Throughout her teaching career she has earned numerous awards such as the *Presidential Award for Excellence in Science Teaching, NMSU Outstanding Research Award, NMSU Research Achievement Award, NASA Trailblazer Award, NASA Innovative Program Award,* the *Disney/McDonald Award, AAAS Larus Award, Who's Who Among America's Teachers,* and special recognition from the New Mexico legislature, as well as being a Fulbright Scholar in

Japan. Brown's research focus is science education and the underrepresentation of minority students and females in the fields of science, math, and engineering. Her grants and publications reflect this focus. She is a National Board Certified teacher, and has facilitated many educators' workshops, presented and published nationally and internationally in peer reviewed journals, in addition to coauthoring two books. The STEM Outreach Center featured in her chapter has thrived for the past 19 years on grant funding. It serves over 5,000 K–8th grade students in both STEM-related afterschool programs and summer programs. Besides serving students, the STEM Outreach Center provides professional development opportunities for K–12th grade STEM teachers.

Vinson Carter is an Assistant Professor at the University of Arkansas, where he teaches courses in Technology & Engineering Education and Elementary Integrated STEM Education. Carter speaks nationally and internationally on STEM education, technology and engineering education, and curriculum development. He is a member of the Executive Board of Directors of the International STEM Education Association (ISEA) and Treasurer of the Council on Technology and Engineering Teacher Education.

Lucia Chacon-Diaz, a native of Chihuahua city, Mexico, is a program coordinator at the STEM Outreach Center for the College of Education at New Mexico State University. She recently graduated with her PhD in Curriculum and Instruction with a concentration in Science Education from NMSU. She also has a Master of Arts in Education, Curriculum and Instruction from NMSU and a Bachelor of Science in Cellular and Molecular Biochemistry from the University of Texas at El Paso (UTEP). In 2017, the same year in which she completed her PhD, Chacon-Diaz was awarded the Hispanic Faculty and Caucus Leadership and Academic Excellence Award. Chacon-Diaz has presented at various national conferences, including the American Chemical Society, American Educational Research Association, the American Association for the Advancement of Science, and the National Science Teachers Association.

Chacon-Diaz started being involved in science teaching during her sophomore year in college, when she was hired by the Chemistry department at UTEP as a Chemistry Peer Leader. In this position, she conducted chemistry workshops for chemistry majors and non-majors. Since college, Chacon-Diaz has taught in several roles at colleges and high schools, including as a health science high school instructor in Mexico.

Yan Carlos Colón is a doctoral candidate in music education at Teachers College, Columbia University. His research interests include democratic education, social justice, and collaborative composing. Born in San Juan, Puerto Rico, Yan studied classical and jazz guitar at the Puerto Rico Music Conservatory earning a BM in Theory & Music Composition and a master's degree in Music Education. In 2016 he obtained an EdM from Teachers College and is currently conducting the study for his dissertation titled: *A Composing Ensemble: Creating Collaboratively with High School Instrumentalists*.As a performer, Yan has worked with artists in Puerto Rico and New York performing many genres of music from Latin and jazz to electronic and indie rock. As a composer, he has written for orchestra, chamber music ensembles and solo instruments. His music has been performed in Puerto Rico, the United States, and Europe. As an educator, Yan developed and teaches a course in music composition at the Teachers College Community School and teaches guitar as a student instructor at Teachers College.

Abiola Farinde-Wu is an assistant professor of urban education in the Department of Leadership in Education at the University of Massachusetts Boston. Her research examines the educational experiences and outcomes of Black women and girls. Highlighting how racial, social, and cultural issues impact the educational opportunities and treatment of Black women and girls, she investigates policies, structures, and practices that influence the recruitment, retention, and matriculation of members of this particular group. Complementing her research, her teaching and service focus on preparing preservice and in-service teachers for diverse student populations. Dr. Farinde-Wu has authored and coauthored numerous studies published in journals, including *Urban Education, Urban Review,* and *Teaching and Teacher Education.* She is also the coeditor of *Black Female Teachers: Diversifying the United States' Teacher Workforce* (Emerald, 2017). She can be reached at abiola.farinde@umb.edu

Aakash Gautum is a graduate student at Virginia Tech. A part of his research concerns collaborating with teachers to design educational technology systems that facilitate their teaching practices with the focus on improving interactions with students.

Aaron J. Griffen, PhD, is the director of diversity, equity and inclusion at DSST Public Schools in Denver, Colorado, and is also an adjunct professor. He is the 2013 recipient of the Urban Teaching Award from Texas A&M University. With over 20 years of experience as an educator, he served as a middle school English teacher, assistant principal in Houston, Texas, and a high school principal in Colorado Springs, Colorado. Dr. Griffen currently serves on the programming committees for the Educating Children of Color Summit and the African American Youth Leadership Conference of Colorado Springs, Colorado, and is chair of School/University Partnerships for the American Educational Research Association's Critical Examination of Race, Ethnicity, Class and Gender in Education Special Interest Group. He is an author, speaker, lecturer, and educational consultant with active research interests and a publishing agenda in diversity, equity, and inclusion, culturally responsive pedagogy, culturally responsive instructional leadership, multicultural curriculum and instruction design and implementation, social justice and advocacy, and urban policy and analysis. He can be reached at Aaron.Griffen@scienceandtech.org

Jesse M. Heines began his teaching career in 1970 at the Anglo-American School in Moscow in the former Soviet Union, where he taught middle school math and science. After graduate school he developed computer-based training courses for Digital Equipment Corporation until 1984. He then joined the faculty in the Dept. of Computer Science at the University of Massachusetts Lowell, from which he retired as a Professor Emeritus in 2016. Jesse's classroom work focused on user interfaces, while his research combined his love of music with computing in a course called "Sound Thinking." That research was supported by three National Science Foundation grants and resulted in *Computational Thinking in Sound,* a book coauthored with his Music Department colleague Gena Greher and published by Oxford University Press (compthinkinsound.org). Jesse now takes courses rather than teaches them, swims regularly, does a variety of volunteer work, and sings his heart out with his barbershop quartet, Fireside. Visit firesidequartet.net and order the CD! For further information, please visit jesseheines.com

Elizabeth Herbert-Wasson is an EdD student in Instructional Technology and Media at Teachers College, Columbia University. She received a BA in Peace, War, and Defense and Linguistics from The University of North Carolina at Chapel Hill, then joined the New York Teaching Fellows Program where she earned an MST in Childhood Special Education from Pace University and spent five years teaching Special Education at a public 6–12 school in the Bronx. While teaching, Elizabeth earned an MEd in Educational Leadership from Baruch College. Elizabeth became an assistant principal at the school in the Bronx. She also served as an assistant principal at a pre-K–8th grade school within Seattle Public Schools. During her time working in schools, Elizabeth developed a passion for incorporating technology and project-based learning into a variety of different curricular areas. She is particularly enthusiastic about exploring participatory pedagogical and research methods with students and to continue to think more deeply about how to cultivate digital proficiencies that are crucial for meaningful citizenship and academic flourishing.

Joseph E. Hibdon, Jr. (Luiseño), PhD, is an Assistant Professor of Mathematics at Northeastern Illinois University (NEIU). Dr. Hibdon came to NEIU while finishing his PhD in Applied Mathematics at Northwestern University where he studied the fluid mechanics of non-premixed flames. Previously, Dr. Hibdon received his bachelor's degree in Mathematics with a minor in Chemistry from the College of the Holy Cross in Worcester, Massachusetts. Currently Dr. Hibdon teaches applied mathematics courses and is the Master of Science in Applied Mathematics advisor for the Mathematics Department at NEIU. His current research interests are in fluid mechanics, dynamical systems, biological modeling, and math education. In particular Dr. Hibdon's recent research has been on ways to get more underrepresented students involved in the STEM fields. Dr. Hibdon is one of the directors for the NEIU MARC U*STAR program where he mentors students on their paths to doctoral programs in the Biomedical Sciences. Dr. Hibdon is also part of two NSF grants on modifying curricula to support STEM learning. One of the NSF grants focuses on integrating research into the curriculum, to increase students' interest in research and help them succeed in their future courses. The goal of the other NSF grant is to add computational thinking to preservice teacher courses, where Dr. Hibdon's role is with a geometry course for future middle school math and science teachers. Dr. Hibdon hopes to continue mentoring students in mathematics and to guide any student who is interested in getting a degree in the STEM fields.

Michael Kerres is full professor of education. He holds a chair of educational media and knowledge management and is the head of Learning Lab at University Duisburg-Essen. Prior to his current appointment, he was professor of educational psychology at Bochum University, and professor of educational technology at Furtwangen University of Applied Sciences. He earned a doctoral degree in psychology at Bochum University and finished his habilitation (postdoctoral thesis) at University of Education in Freiburg. His present research interests include digital transformation of education, instructional designs for digital learning and learning infrastructures.

Hanna Kim, PhD, is an Associate Professor at Northeastern Illinois University. Dr. Kim received her PhD in Science Education from the University of Texas at Austin while working at the Bureau of Economic Geology as a research assistant, where she was involved with state-funded oil and gas projects such as STARR (State of Texas Advanced Resource Recovery). Her research focuses on integrating interactive simulations and computational thinking in elementary and middle school science teaching and learning. As principal investigator of InSTEP: Inquiry-Based Science and Technology Program, she collaborated with science teachers in six Chicago Public Schools to help diverse students learn science in a fun way using innovative technology and an inquiry approach. Her publications in the *Journal of Science Education and Technology* and *Journal of Research and Practice* support this endeavor. She has taught elementary and middle school science methods, science, and math connectivity, student teaching seminars, clinical supervision, and research methods for teachers. As senior personnel of the NSF STEM+C grant, she investigates how coding and robotics affect preservice teachers' computational thinking skills, which can influence their future students in science classrooms. Her preliminary study in computational thinking for preservice teachers was published in *Education and Information Technologies*. Currently, she is working with preservice teachers in her science methods course as they use earthquake simulator robotics with Lego Mindstorm software. The results of this robotics study will help the STEM+C faculty team to redesign our elementary and middle school science curriculum to embed computational thinking

Carol-Ann Lane completed her Doctor of Philosophy at the Faculty of Education at the University of Western Ontario in February 2018. Her doctoral research focused on Curriculum Studies—specifically exploring boys' meaning-making and the cultural knowledge they gained through their video gaming practices. Her interest in this field stems from her own experiences and dedication as a qualified Ontario teacher working for the Dufferin Catholic Peel District School Board in the primary Grade 3 division, junior division, and intermediate/senior division, and as a teaching associate for the bachelor of education and graduate level courses (including online). She has been an owner of a tutoring company (for which she is also an instructor) for more than six years. Lane is deeply committed to examining the potential of secondary learning domains, such as video games, for in-school learning processes. Informing her research is the multiliteracies framework and learning by design theories (Cope & Kalantzis, 2009, 2016) Through her multicase ethnographic research study on which her chapter is based, she critically considered the nature of understudied boys' out-of-school video gaming practices. Past research has focused on the themes of violence and misogyny associated with video games. However, there is a misunderstanding of the embedded narratives that exist in video games which needs further attention. Her research has the potential to inform educators on ways boys navigate their selection of games and develop, adopt, and share their cultural knowledge through various modes of meanings.

Ahlam Lee has served as an assistant professor of leadership studies doctoral program at Xavier University since 2015. Prior to that, she had served as an assistant professor of education at Arkansas State University for three years, after having completed her postdoctoral training in the Graduate School of Education at University of Pennsylvania in 2012 and receiving her PhD in educational leadership and policy analysis from the University of Wisconsin-Madison in 2011. She earned her master's degree in rehabilitation psychology at the University of Wisconsin-Madison in 2008, another master's degree in public finance and budgeting at Columbia University in 2005, and a bachelor's degree in business economics and public policy at the Indiana University-Bloomington in 2001. Lee's interdisciplinary background enables her to be involved in various research areas including STEM education, refugee resettlement issues, and transformative leadership. In the field of STEM education research, she has published her works in top-ranked journals such as Computers & Education and Teachers College Record. In terms of refugee resettlement issues, she published North Korean Defectors in a New and Competitive Society: Issues and Challenges in Resettlement, Adjustment, and the Learning Process (Lexington Books, 2015), and her book chapter entitled *The Korean Wave: A Pull Factor for North Korean Migration* is forthcoming (Routledge, in press). In terms of research on transformative leadership, she is developing a comprehensive model of hierarchical microaggressions in higher education settings, which is funded by the National Center for Institutional Diversity.

Kemper Lipscomb is a PhD candidate in STEM education at the University of Texas at Austin. Her research examines how students investigate and explain scientific phenomenon. Her current work focuses on students› participation in modeling, computational thinking, and argumentation practices in the context of collaborative scientific investigations. She has published articles and books relating to these issues. She assists in teaching science and science education courses at the undergraduate level, with a teaching philosophy rooted in theories of learning that stress the importance of shared experience, collaboration, and metacognition.

Susan Lowes, PhD, is Director of Research and Evaluation at the Institute for Learning Technologies at Teachers College, Columbia University. She has conducted research at both the university and K–12 levels, with a focus on the impact of technology on teaching and learning, and directed evaluations of multiyear projects funded by the U.S. Dept. of Education, the National Science Foundation, state and local departments of education, and private foundations. Her research focusses on how technology can be used to teach STEM concepts and on online learning at the K–12 level. She is also Adjunct Professor in the Program in Computers, Media, and Learning Technologies Design at Teachers College, teaching a course on Researching Technology in Educational Environments and a course on Online Schools and Online Schooling for K–12. She received her PhD in Anthropology from Columbia University for work on the island of Antigua in the West Indies and still does some research there when she has some spare time.

Scott Mayle, MS, is an Instructor in the Department of Physics and Astronomy at Northeastern Illinois University (NEIU). Mr. Mayle received a Master of Science degree from Northwestern University, where he studied nano-circuits made from novel materials. While at Northwestern University, he was awarded the NSF GK12 fellowship. This fellowship allowed him the opportunity to work with local teachers to bring modern research topics and computational thinking techniques into their classrooms. Mr. Mayle was recruited into the NEIU Physics and Astronomy Department to design and teach a circuits course and lab. Since then, he has taken on an expanded teaching role. He teaches the university physics course with an emphasis on research and also teaches the introductory physics class for the Math, Science and Technology for Quality Education (MSTQE) program. The MSTQE program prepares future middle school teachers to teach science and math, and to integrate computational thinking into their teaching practices. Mr. Mayle is currently serving as senior personnel on an NSF STEM+C grant, with a focus on bringing computation into the core content classes of the MSTQE program. In addition to teaching, Scott runs an interdisciplinary seminar for a student peer-leader program at NEIU. This seminar covers a variety of topics suited for students in the STEM fields, including facilitating group work, experimental design, ethics in the STEM fields, writing skills, presentation skills, and the peer review process.

R. Martin Reardon, PhD, is an Assistant Professor in the Department of Educational Leadership within the College of Education (COE) at East Carolina University (ECU). He is currently a coprincipal investigator on a National Science Foundation grant focused on the integration of computational thinking with the teaching of music and visual arts in rural middle schools which is completing its first full year of operation in fall 2018. This is the involvement which gave rise to the chapter in this volume which he coauthored with one of his graduate assistants, Claire Webb. He also has a research interest in the implementation of trauma-sensitive approaches in rural schools as an associate of the Rural Education Institute within the ECU COE. He is a founding editor (together with Jack Leonard) of the *Current Perspectives on School/University/Community Research* book series (this is the fourth volume), which he initiated during his second term as the Chair of the eponymous American Educational Research Association' Special Interest Group from which the motivation to establish the *Current Perspectives* series emerged.

Jennifer E. Slate, PhD, is a Professor of Biology at Northeastern Illinois University (NEIU). After receiving her PhD from the University of Louisville, Dr. Slate completed a postdoctoral fellowship at the University of Michigan. At NEIU, she teaches a wide variety of courses from the introductory level to the graduate level, including a biology course for preservice elementary and middle school teachers. Dr. Slate has extensive curriculum development experience in designing inquiry-based lab and field exercises for both general education and upper-level courses, resulting in five competitive Teaching Excellence Awards (2017, 2014, 2011, 2009, and 2006). As a coprincipal investigator on an NSF STEM+C grant, she is developing computational thinking activities that allow preservice teachers to use modeling to explore complex biological systems. Dr. Slate is passionate about computational thinking in K–12 education, having coached a team of middle school girls as they designed and programmed robots and researched STEM topics for the First Lego League (http://www.firstlegoleague.org/) robotics competition. To help others involve their students in computation-based learning, Dr. Slate has given multiple workshops to college faculty and to high school teachers about integrating authentic research experiences, data analysis, and modeling into their courses. She regularly mentors the independent research projects of undergraduate students, over thirty of whom have presented their research results at regional and national scientific conferences. To interest and engage the public with science, she gives presentations and nature tours to the community at venues such as Indiana Dunes National Lakeshore and Volo Bog Nature Preserve.

Elizabeth E. Smith is Assistant Professor of Education at the University of Tulsa and serves as the Chair of the Department of Education. Prior to this position, she led a variety of school-university partnerships for more than a decade. In addition to studying P–20 partnerships, she conducts research on assessment in higher education and the policy contexts of P–12 school funding.

Sudha Srinivas, PhD, is a Professor of Physics at Northeastern Illinois University (NEIU). Dr. Srinivas received her PhD in Theoretical Condensed Matter Physics from the University at Albany, State University of New York, where she studied the electronic structure and hyperfine properties of high temperature superconductors. This was followed by postdoctoral research on the computational modeling of metallic nano-clusters at Argonne National Laboratory, Argonne, IL. Dr. Srinivas has been at NEIU since 2005 and has taught courses ranging from introductory physics to graduate-level quantum mechanics and condensed matter physics. Additionally, she has taught within the Math, Science and Technology for Quality Education (MSTQE) program which prepares future middle school teachers and within the University Honors Program. Dr. Srinivas's scholarly interests include first-principles modeling of the electronic and magnetic properties of condensed matter systems and working on strategies to better engage students in the STEM fields. She has served as principal investigator/coprincipal investigator on several external grants focused on materials theory, STEM education, and student engagement and high-impact pedagogical practices in STEM education. As a co-principal investigator of the NSF STEM+C grant focused on the Biology, Math, Physics, and Computer Science content of the middle school math and science preservice teacher education program at NEIU, Dr. Srinivas is working with faculty from these departments on embedding computational thinking into math and science courses. Dr. Srinivas currently serves as the Acting Associate Dean of the College of Arts and Sciences.

Nancy Streim, PhD, is Associate Vice President for School and Community Partnerships at Teachers College, and Special Advisor to the Columbia University Provost. She earned a bachelor's degree with honors from Bryn Mawr College, a master's degree from the State University of New York, and a PhD from the University of Wisconsin-Madison. Prior to her role at Teachers College, she served as Associate Dean for Educational Practice at the University of Pennsylvania's Graduate School of Education. Dr. Streim leads TC's efforts to apply university expertise to improving public education for New York City's children. She is an architect of a "university-assisted community schools model" that systematically addresses conditions related to educational success, including teacher development, expanded learning opportunities, physical and mental health, and parent engagement. She has established two successful university-assisted public elementary schools—one in Philadelphia and one in New York City—and has built long term partnerships with local education authorities, community organizations and the corporate sector. Dr. Streim has been Principal Investigator on grants and contracts from the National Science Foundation, New York State and New York City departments of education, General Electric Foundation and JPMorgan Chase Foundation among others. She also manages College-wide fellowship programs that provide financial aid for students to carry out teaching, research, and service with local schools and organizations.

Woonhee Sung received her EdD in Instructional Technology and Media at Teachers College, Columbia University. She also has an MA in Instructional Technology from the University of Texas at Austin and another in Applied Statistics from Teachers College. Her current research focusses on the design, development, and implementation of technology-integrated learning environments that foster computational thinking and integrate the engineering design process. At Teachers College Community School, she has also designed and assessed a curriculum that combines the concept of programming with mathematics concepts (number line, number sense, arithmetic, second geometry) and another that combines robotics with science concepts (axles, wheels, gears).

Deborah Tatar (PhD Psychology, Stanford) is Professor of Computer Science and, by courtesy, Psychology at Virginia Tech as well as a fellow of the Center for Creativity, Art, and Technology and a Member of the Program for Women and Gender Studies. Previously, she was a Cognitive Scientist at the Center for Technology in Learning at SRI International, a member of the research staff at Xerox PARC and a Senior Software Engineer at Digital Equipment Corporation. Her research is concerned with the integration of technology into face-to-face contexts, especially middle school classrooms. Her focus has been on mathematics and science but always prioritizes interpersonal interaction and the ability of the teacher to operate effectively in the environment. She was the Principal Investigator on the CHEM+C project.

Lalitha Vasudevan holds a BA in Psychology and a PhD in Education from the University of Pennsylvania. She is Professor of Technology and Education in the Communication, Media, and Learning Technologies Design Program at Teachers College, Columbia University. Over the past 20 years, she has explored the intersection of adolescent literacies, media and technologies, youth culture, and juvenile justice. She engages participatory, ethnographic, and multimodal methodologies to study how youth craft stories, represent themselves, and enact ways of knowing through their engagement with literacies, technologies, and media. Lalitha has conducted research with court-involved youth including: a longitudinal, ethnographic study in an alternative to incarceration program; an oral history based qualitative research project with young men at Rikers Island; and an ongoing multisited study of participatory, arts-based, multimedia storytelling with adolescents at an afterschool program located in an alternative to detention program. She has also explored the pedagogical practices of inclusive and special education teachers, the literacy and identity practices of middle school adolescents inside classroom settings, and the multimodal literacy and media engagements of adolescent boys. Lalitha has coedited two volumes that explore the intersections of youth, media, and education: *Media, Learning, and Sites of Possibility* and *Arts, Media, and Justice: Multimodal Explorations with Youth* (both published with Peter Lang), and is currently writing a book about education, multimodal play, and belonging in the lives of court-involved youth.

Bettina Waffner studied political science and history as well as peace and conflict studies at the FernUniversität in Hagen. In 2016 she earned her doctorate with a thesis on decision-making structures in the European Council. From 2013 to 2016, Waffner was research assistant at the Chair of International Politics at the FernUniversität in Hagen. She has been research assistant at the Chair of Educational Media and Knowledge Management (Learning Lab) at the University of Duisburg-Essen since the end of 2016. She heads the team "School in digital change." Her work focuses on testing and researching innovative teaching and learning scenarios that support cooperative, collaborative, and participatory work with digital media.

Daniel A. Walzer is an assistant professor of music at the University of Massachusetts Lowell. Walzer received his PhD in Leadership from the University of the Cumberlands, his MFA from Academy of Art University, his MM from the University of Cincinnati, and his BM from Bowling Green State University. Originally trained as a percussionist, Walzer maintains an active career as a composer and music technology researcher. For more information, please visit http://www.danielwalzer.com

Claire Davie Webb has her bachelor's degree in psychology from Asbury University, and is currently obtaining her master's in Marriage and Family Therapy from East Carolina University. She is presently a therapist-in-training at the Marriage and Family Therapy clinic on campus and works with individuals, couples, and families. Claire is also currently gaining experience as a therapist in several schools across eastern North Carolina, working with students with behavioral problems, family struggles, or difficulties in school. Claire will graduate in May and hopes to work with populations experiencing grief or loss.

Heather D. Young is an Assistant Professor in the Department of Curriculum and Instruction at the University of Arkansas. She currently serves as the program coordinator for the Childhood and Elementary Education programs and as the Director of the University of Arkansas Clinic for Literacy. Dr. Young is a former public school kindergarten teacher and literacy specialist. Her research interests include literacy education and assessment, the development of professional dispositions, and the impact of authentic and intensive experiences on teacher preparation.

Printed in the United States
By Bookmasters